Advances in Bioelectrochemistry Volume 4

Frank N. Crespilho
Editor

Advances in Bioelectrochemistry Volume 4

Biodevice, Bioelectrosynthesis and Bioenergy

Springer

Editor
Frank N. Crespilho
University of Sao Paulo
São Carlos, São Paulo, Brazil

ISBN 978-3-030-99661-1 ISBN 978-3-030-99662-8 (eBook)
https://doi.org/10.1007/978-3-030-99662-8

This Springer imprint is published by the registered company Springer Nature Switzerland AG
The registered company address is: Gewerbestrasse 11, 6330 Cham, Switzerland

Contents

Protein Engineering for Designing Efficient Bioelectrodes

Andressa Ribeiro Pereira

Abstract Protein engineering gives interesting solutions for the development of efficient bioelectrodes, once it possibilities the enhancement of the electrochemical performance of bioelectronics and improves the properties of the redox proteins. Thus, in this chapter how the biosensors and biofuel cells work was discussed briefly; the electron transfer catalyzed by enzymes in a direct and a mediated system was explained. However, the focus here is the modification of the protein structure to improve the performance of the bioelectrodes using protein engineering. The main methods that are utilized to change the redox protein structures, such as rational design and directed evolution, were explained. Moreover, some examples of bioelectrodes, which had their performance improved after the modification, were shown.

In the last years, electrochemistry has been utilized to understand the biochemistry of proteins, for instance, enzymes, besides being a powerful method to explore these biomolecules in biosensors and bioelectronics. As known, proteins play an important role in bioelectronic devices. For bioelectrochemical sensors, for example, there is a transduction of a biomolecular recognition event into an electrical signal [1, 2]. Many of these biosensors are based on the amperometric principle, where the recognition of the analyte is in a biomolecular way and is coupled to an oxidation or reduction process, which gives rise to an electrical current [3]. Although there are other sensor types, the amperometric ones are very relevant ones; their enzyme-coated electrodes are similar to the function and architecture of biofuel cells.

Biofuel cells are used to convert the chemical energy of organic molecules into electrical power, and have attracted significant attention because the electrical energy obtained is clean and portable [3]. In the specific case of the enzymatic biofuel cells (Fig. 1), the catalysts are enzymes that are immobilized on the surface of the electrodes.

A. R. Pereira (✉)
São Carlos Institute of Physics, University of São Paulo, São Carlos, São Paulo, Brazil
e-mail: andressa.arp@gmail.com

F. N. Crespilho (ed.), *Advances in Bioelectrochemistry Volume 4*,
https://doi.org/10.1007/978-3-030-99662-8_1

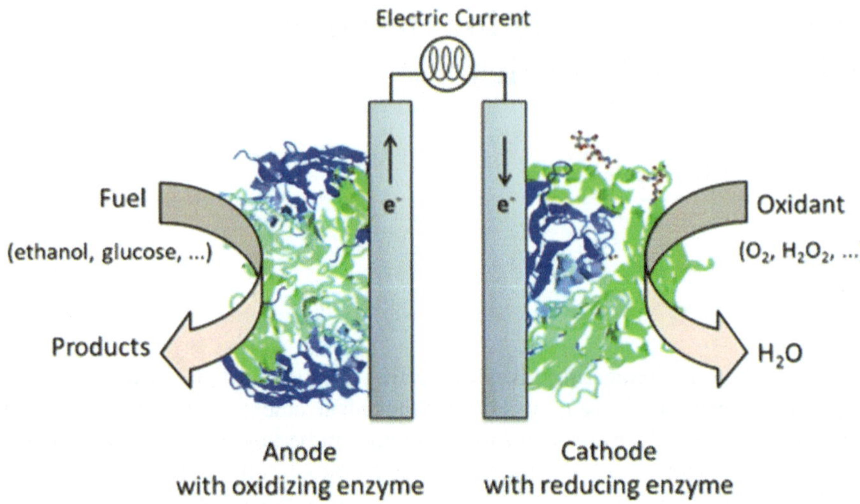

Fig. 1 Representative scheme of an enzyme-based biofuel cell. Reproduced with permission from Ref. [4]. Copyright 2021 John Willey and Sons

Although the utilization of these biomolecules is promising, some issues need to be overcome. The utilization of enzymes, for example, has many advantages, such as biocompatibility, high selectivity and catalytic efficiency at physiological conditions, but presents some concerns, since they have poor conductivity and low-volumetric catalytic activity, which results in low current density [5]. Another issue that deserves to be mentioned is the limited distance of the enzyme cofactor and the electrode surface, which could be the limiting factor for the success of the biodevice [6] in the non-mediated systems.

Thus, thinking about improving the performance of the non-mediated systems, a possibility that has been used in the last years is the modification of the electrode surfaces, such as carbon nanotubes [7, 8], graphene [9, 10] or the incorporation of the nanoparticles to the electrodes [11–14]. More recently, another option has been utilized to improve biodevices: the modification of the protein structure, which is known as protein design and protein engineering. In addition to improving biodevices, protein engineering allows understanding protein-electrode interactions, and the structure and function of redox proteins aiming at structural stability and functional capabilities [15, 16].

Therefore, this chapter aims to show how the electron transfer occurs in the redox proteins, how redox proteins could be utilized in bioelectrodes and the types of protein engineering that could be utilized to improve the efficiency of the biodevices.

1 Electron Transfer Catalyzed by Enzymes

Enzymes are proteins known for catalyzing reactions in the biological system. They have extraordinary catalytic power, high degree of specificity, and work at mild temperature and pH conditions. They are classified according to the reactions that they catalyze, and although there are six classes of enzymes, here just one is important: oxidoreductase, which is responsible to transfer electrons [17].

Because of the property of electron-transfer reactions between electrode surfaces and enzymes, redox proteins have been explored over the last years in order to develop efficient biosensors and biofuel cells, since the electron transfer plays a central role in biological energy conversion, for instance, in the respiratory chain and in the citric acid cycle [18]. In general, enzymatic reactions that occur on the electrode surface can proceed in two ways. The first one is based on the utilization of redox mediators (Fig. 2a), where the enzyme catalyzes the oxidation and the reduction of the mediator [19]. In this kind of system, the catalytic process involves the modification of the analyte and the mediator, and the utilization of these species is needed due to the bulky protein structure of some enzymes, which distances the cofactor from the electrode surface. The second one corresponds to the direct electron transfer (DET) [20] (Fig. 2b). In this case, the electron is transferred directly from the active center of the enzyme to the electrode surface, without the help of any other species, providing some important information in relation to thermodynamics and kinetics of the biological process.

The electron transfer between the active center of an enzyme and the electrode surface is very important to the development of biodevices [21], since the efficiency of the transfer and the immobilization methods of the enzymes influence the performance of the bioelectrodes. Some conditions need to be satisfied for the occurrence of electron transfer in redox proteins: according to Marcus theory [22], the electron transfer rate between two redox centers will depend on three factors: reorganization

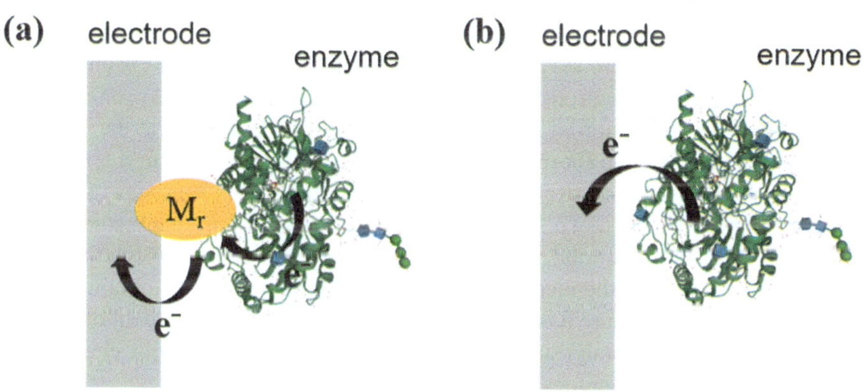

Fig. 2 Schematic representation of **a** mediated electron transfer, where M_r corresponds to the redox mediator, and **b** direct electron transfer

energy, divided in the intern and the extern spheres contributions, which corresponds to the solvent reorientation due to the modification of the molecule and its charges, and the energy that is necessary to modify the distances of the bonds [23], the potential difference between redox centers and the distance between these center [24]. Thus, electron transfer has been observed in redox proteins where the active center is located near the surface of the protein. However, the challenge is related to obtaining the electron transfer for enzymes which the active center is located deep in the protein cavity, where access is difficult. In this case, electron transfer between the active center and the electrode is slow, although attractive due to the simplicity of the system [20, 25]; beyond the fact that this type of electrode could work at a potential range close to the redox potential of the enzyme, there is a challenge to obtain this electron transfer efficiently, since in some cases, as for glucose oxidase (GOx), for instance, its redox center—FAD—is deeply inside in the protein core [26, 27], at a distance of 13–15 Å of the surface [28].

Therefore, aiming to promote an efficient electron transfer, some structural modifications could be done in the protein to improve the electrochemical response between protein and the electrode surface, as will be shown in topic 16.3, with protein engineering.

2 Bioelectrocatalysis Employing Redox Enzymes

The utilization of alternative energy supply sources is a very important topic nowadays and miniaturized biofuel cells could be used for this purpose in the future as nano-microelectronic devices and biosensors [29]. One of the advantages of biofuel cells is that they are able to convert the chemical energy of the biological environment into electrical current. Two approaches are possible: the utilization of microorganisms and/or enzymes as biological reactors to do the fermentation of the raw materials to fuel products [30], and the utilization of the microorganisms and/or enzymes as catalysts, which converts chemical energy to electrical one [31]. Although the different possibilities, here the focus is in the systems that utilizes redox enzymes to develop promising bioelectrodes, once the application of purified redox enzymes for the oxidation and reduction of specific fuel and oxidizer substrates at the electrode surfaces is more effective for the development of biofuel cells. The first application of enzymes as biocatalysts occurred in 1970 even though the properties of bioelectrocatalysis have been known for a long time [32]. Since then, many studies have been done to develop immobilization strategies aiming to orientate the active site of the biocatalyst close to the electrode, since this is a crucial step to obtain efficient devices. Thus, the literature has many works describing the development of systems based on DET in the analytical device by protein modification with chemical or genetic engineering techniques [33–35].

For the occurrence of direct bioelectrocatalysis, the DET mechanism, which corresponds to the electron transfer between the electrode surface and enzyme, occurs directly in the absence of any mediator. Besides, for the occurrence of DET some specific conditions need to be satisfied, as follows: there is a minimal distance between the electron donor—cofactor—and acceptor in order for the fast electron transfer to occur and to improve bioelectrocatalysis, and the corrected orientation of the enzyme (with the cofactor) is immobilized onto the electrode surface to maintain the electron tunneling distance below 2 nm [36]. Nevertheless, sometimes, the successful orientation and the small distance between the redox cofactor and the electrode surface is not enough for DET, since the orientation could limit the access to the enzyme preventing the success of bioelectrocatalysis [37].

In relation to the application of bioelectrocatalysis, it is important to mention that biosensors are utilized in many fields, such as environmental and medical science and in food/agriculture depending on the kind of target analytes, and the enzymes are very important once they are highly specific biological catalysts, which detect low concentrations of target agents without interference. Besides, these systems have the advantage of being low cost, fast response and enables the application in portable devices without preparing samples [32]. Biomedical sensors are specific types of biosensors, which provide the necessary interface between the biological material and the electrode surface to detect medically relevant parameters. For instance, GOx was the first enzyme that was utilized in medical biosensors for glucose monitoring in diabetic patients [38]. Another application of these biomedical sensors are in cancer diagnosis; since they reduce the assay time, they are portable, have high sensitivity and selectivity, and have the possibility to be miniaturized. For example, genosensors could be utilized to detect prostate cancer using electrochemical methods and optical absorption. In this case, it is used a biomarker, referred to as PCA3 [39], a prostate cancer gene 3.

Environmental sensors are another type of biosensors that have been developed lately. With the increasing growth of environmental pollution, the development of a sensitive and innovative technique to control this issue is important. For instance, the acetylcholine esterase enzyme has been utilized to build biosensors for the detection of carbamate pesticides and organophosphorus, since these compounds inhibit the activity of the acetylcholine esterase [32]. Another important area that accurate analysis is essential is in the food and drink products, where biosensors have been used to determine glucose and some other sugars [40–42]. Besides, they could be utilized for the identification of bacterial contamination in food [43].

3 Protein Engineering

Protein engineering has gained prominence because it is an important tool to understand the effect of the properties of the biomolecules on device performance, since it allows precise control over molecular architecture [5]. It offers interesting solutions for the development of bioelectrodes, enhances the electrochemical performance of bioelectronics and improves the properties of the redox proteins. Besides, with this strategy, the sensitivity, selectivity and power density could be improved.

One of the approaches of protein engineering is to genetically modify the enzymes to approximate them to nanoparticles. For instance, GOx was modified to display a free thiol group near the active site [44]; although this strategy decreased the catalytic activity of the enzyme, direct communication with the nanoparticles and the electrode surface was obtained. Another possibility to utilize protein engineering is to organized metal structures while simultaneously linking enzymes to these structures. Once again, GOx was utilized, and it was fused to a ring-shaped protein with functional domains that bind gold nanoparticles via adsorption in the inner pore [45]. In this case, the design allows to building layers of electrically connected enzymes to the electrode surface, facilitating the electronic wiring between multiple functional enzymes, once the 3D organization is facilitated. Thus, this strategy of protein engineering is promising since it protects enzymes from denaturation, increases the enzyme-based electrode performances and approximates the reactive elements to their active sites [5].

The main approach that protein engineering utilizes includes rational design, directed evolution and combined effects. They have been used to improve the utilization of biocatalysts at different conditions and applications, such as tolerance toward organic solvents, peroxide resistance [46] and use of non-natural cofactors [47, 48]. Rational design is the method that utilizes site-directed mutagenesis, which verifies the protein-function relationships. In this case, the computer models and the hypotheses of the researchers are based on available crystal structures of the protein, and the property that needs to be improved [49]. Rational design is an approach universally applicable, which is not limited to specific protein folds or motifs. Besides, the new computer technologies allow the rapid screening of a lot of data without modeling individual structures [16, 50]. Among the strategies used in rational design could be cited (i) trimming the "fat", where the amino acids that are not essential for the protein function and those to maintain the structure are removed, and the active site is approximated to the electrode surface; and (ii) protein surface modifications, such as deglycosilation or the oligomerization of the enzyme.

In protein oligomerization, for example, two or more polypeptide chains are involved in dynamic interactions [51]. The advantages of this process are the allosteric regulation and the enhanced resistance to denaturation and degradation for large proteins [52, 53]. Protein oligomerization is considered as a type of protein aggregation, and depends on extrinsic and intrinsic factors associated with protein environment and protein structure levels [54]. For instance, the development of GOx oligomers (Ol-GOx) via chemical route with a Brønsted acid could be cited [35], where the exposure of the hydrophobic chains with the Brøonsted acid was observed, allowing the oligomerization of GOx. After the procedure, the bioelectrochemistry of Ol-GOx was studied and a maximum rate constant for heterogeneous electron transfer sevenfold higher than that for native GOx was observed, as illustrated in Fig. 3.

In directed evolution, it is not necessary to know the structures of the enzymes; in this process, it is important to discover the fundamental design principles of electron transfer in redox proteins. This strategy utilizes the power of Darwinian selection in the evolution of proteins with interesting properties that are not found

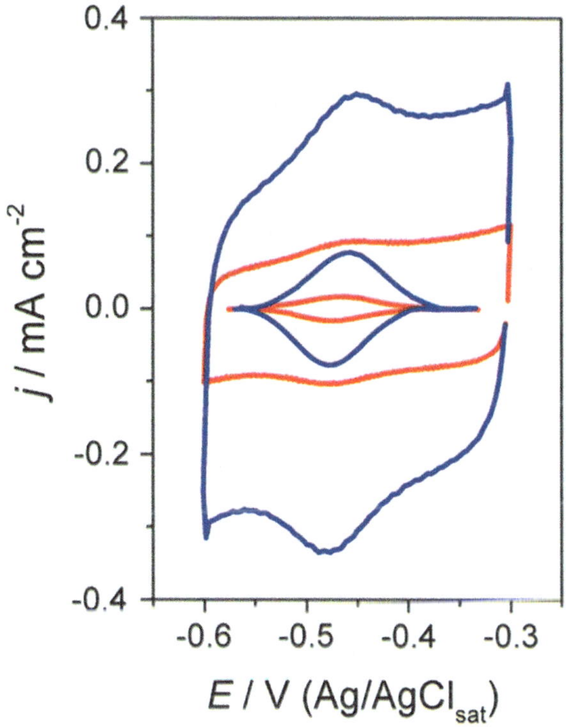

Fig. 3 Cyclic voltammograms of GOx bioelectrodes (red line) and Ol-GOx bioelectrodes (blue line). Conditions: scan rate: 100 mV s^{-1}; electrolyte support: 0.1 mol L^{-1} sodium phosphate buffer (pH 7.5); Temperature: 25 °C; Argon atmosphere. *Inside*: cyclic voltammograms after the subtraction of the capacitive current. Reproduced with permission from Ref. [35]. Copyright 2021 ACS Publications

in nature [55]. However, the experiments in this case are significantly more time-consuming than in rational design. Directed evolution has been utilized and well accepted in academia and in industrial biotechnology, besides, it could be used by computational biologists to elucidate structure-functions relationships for proteins with known crystal structures [49]. Moreover, this method is important to discover the fundamental design principles of electron transfer in redox proteins when proteins with known structures are utilized [16]. In directed evolution, some different mutated enzymes via DNA modification are created, and resulting proteins are identified by selection or screening. One of the most methods used for creating a diversity of the genetic level is random mutagenesis by error-prone polymerase chain reaction (epPCR), however, one alternative method for this one for generating genetic diversity is combining diverse genes that were selected for function by natural evolution [56]. For instance, by directed evolution, the mediated activity of a monooxygenase, P450 BM-3 from *Bacillus megaterium*, was improved [57], and the variant contains five amino acid substitutions. It was described that P450 BM-3 improves twice the activity for an alternative cofactor system in which NADPH has been replaced by the cobalt(III)sepulchrate mediator and Zn dust as electron source. Another example of successful directed evolution was the improvement in the thermal stability of glucose dehydrogenase from *Bacillus megaterium* by introducing two amino acid substitutions [58].

Domain shuffling, also known as molecular Lego, has been used as a genetic engineering method to fuse or swap functional modules in some proteins aiming to mimic the natural molecular evolution in redox protein engineering. In bioelectrochemistry, this method has been utilized as a prototype for engineered systems and it could be utilized an engineered disulfide bridge or a genetically engineered peptide linker between two proteins as a link between the protein domains [16]. Another methodology has been used is the Protein Reconstitution, which examines the structure-function relationship of proteins. In this method, the native redox-active center of the protein is eliminated and after that, another related cofactor or ion is implanted in the reconstituted protein, which aims the development of new functions that are absent in the native proteins [16].

One of the possibilities of protein engineering is to improve the specificity and the sensitivity of the enzymes. For acetylcholine esterase, which hydrolyzes acetylcholine, there is a study which employs molecular modeling of the active site in combination with targeted substitution of amino acids [59] to make an enzyme with a very lower detection for the analyte dichlorvos [60]. For enzymatic biofuel cells, the efficient conversion of the available biomolecule types into electronic current is important, however, complete oxidation of all molecules is difficult to achieve. Because of that, many strategies could be utilized to solve this problem, and one of them is the engineering of the isolated enzymes, as cited before; moreover, can also be use the development of native multi-enzyme systems [61], which are systems that utilized biological enzymes cascades, such as in the citric acid cycle.

Increasing the sensitivity in sensor and the power density in biofuel cells is very important and to do that, current density is an important parameter. Thus, to improve this parameter, some knowledge in rational design is required. It is important to mention that electron transfer could be improved in non-protein and protein-based approaches [62–65], but here the non-protein approaches are out of the scope. In relation to protein-based strategies, as cited before, some proteins are capable of DET [36, 66], while others have some difficulties, such as the localization of the active center. Therefore, protein engineering could solve this limitation by changing the structure of the proteins aiming to improve the access to the active center and/or promoting the correct orientation of the protein to the electrode surface [67]. Usually, this approach reduces the distance in which the electrons need to tunnel. The last examples show that protein engineering has been utilized to improve functional performance, where these approaches utilize directed mutagenesis to adjust the biomolecular recognition properties or to directly immobilize the protein onto the electrode surface.

4 Conclusions

As demonstrated, the utilization of redox proteins is promising for the development of biosensors and biofuel cells, and the utilization of these biomolecules has many advantages, such as biocompatibility, high selectivity and catalytic efficiency at physiological conditions. However, some issues need to be overcome to improve the performance of the biodevices. In this way, in the last years some researchers have studied the modification of the electrode surfaces or the modification of the protein structure, which is known as protein design or protein engineering. One of the main advantages of protein engineering is the possibility to understand the effect of the properties of biomolecules on device performance, enhancing the electrochemical performance of bioelectronics and improving the properties of redox proteins. There are many methodologies of protein engineering, but the two most used are rational design and directed evolution. The first one utilizes site-directed mutagenesis, where the computer models and the hypotheses of the researchers are based on available crystal structures, and the method utilizes the power of Darwinian selection in the evolution of proteins. Protein engineering has been seen as an important tool to improve the specificity and the sensitivity of the enzymes, which is important in the development of bioelectrodes for biosensors and biofuel cells.

References

1. Grieshaber, D., MacKenzie, R., Voros, J., Reimhult, E.: Electrochemical biosensors—Sensor principles and architectures. Sensors **8**, 1400–1458 (2008)
2. Willner, I., Katz, E.: Bioelectronics: from Theory to Applications. Willey VHC2005
3. Caruana, D.J., Howorka, S.: Biosensors and biofuel cells with engineered proteins. Mol. BioSyst. **6**, 1548–1556 (2010)
4. Luz, R.A.S., Pereira, A.R., de Souza, J.C.P., Sales, F.C.P.F., Crespilho, F.N.: Enzyme biofuel cells: thermodynamics kinetics and challenges in applicability. Chemelectrochem **1**, 1751–1777 (2014)
5. Renner, J.N., Minteer, S.D.: The use of engineered protein materials in electrochemical devices. Exp. Biol. Med. **241**, 980–985 (2016)
6. Degani, Y., Heller, A.: Direct electrical communication between chemically modified enzymes and metal-electrodes. 2. Methods for bonding electron-transfer relays to glucose-oxidase and D-amino acid oxidase. J. Am. Chem. Soc. **110**, 2615–2620 (1988)
7. Azamian, B.R., Davis, J.J., Coleman, K.S., Bagshaw, C.B., Green, M.L.H.: Bioelectrochemical single-walled carbon nanotubes. J. Am. Chem. Soc. **124**, 12664–12665 (2002)
8. Zhao, H.Y., Zheng, W., Meng, Z.X., Zhou, H.M., Xu, X.X., Li, Z., Zheng, Y.F.: Bioelectrochemistry of hemoglobin immobilized on a sodium alginate-multiwall carbon nanotubes composite film. Biosens. Bioelectron. **24**, 2352–2357 (2009)
9. Gao, H., Duan, H.: 2D and 3D graphene materials: preparation and bioelectrochemical applications **65**, 404–419 (2015)
10. Kuila, T., Bose, S., Khanra, P., Mishra, A.K., Kim, N.H., Lee, J.H.: Recent advances in graphene-based biosensors. Biosens. Bioelectron. **26**, 1637–1648 (2011)
11. Katz, E., Willner, I.: Integrated nanoparticle-biomolecule hybrid systems: synthesis, properties, and applications. Angew. Chem. Int. Ed. **43**, 6042–6108 (2004)
12. Xiao, Y., Patolsky, F., Katz, E., Hainfeld, J.F., Willner, I.: "Plugging into enzymes": nanowiring of redox enzymes by a gold nanoparticle. Science **299**, 1877–1881 (2003)
13. Crespilho, F.N., Ghica, M.E., Gouveia-Caridade, C., Oliveira, O.N., Brett, C.M.A.: Enzyme immobilisation on electroactive nanostructured membranes (ENM): optimised architectures for biosensing. Talanta **76**, 922–928 (2008)
14. Pereira, A.R., Iost, R.M., Martins, M.V.A., Yokomizo, C.H., da Silva, W.C., Nantes, I.L., Crespilho, F.N.: Molecular interactions and structure of a supramolecular arrangement of glucose oxidase and palladium nanoparticles. Phys. Chem. Chem. Phys. **13**, 12155–12162 (2011)

15. Davis, J.J., Hill, H.A.O., Bond, A.M.: The application of electrochemical scanning probe microscopy to the interpretation of metalloprotein voltammetry. Coord. Chem. Rev. **200**, 411–442 (2000)
16. Prabhulkar, S., Tian, H., Wang, X.T., Zhu, J.J., Li, C.Z.: Engineered proteins: redox properties and their applications. Antioxid. Redox Signal. **17**, 1796–1822 (2012)
17. Nelson, D.L., Cox, M.M.: Principles of Biochemistry, 4th edn. Freeman and Company, New York (2005)
18. Scheller, F.W., Wollenberger, U., Lei, C., Jin, W., Ge, B., Lehmann, C., Lisdat, F., Fridman, V.: Biolectrocatalysis by redox enzymes at modified electrodes. Rev Molecul Biotechnol **82**, 411–424 (2002)
19. Cardosi, M.F., Turner, A.P.F.: Biosensors: fundamentals and Applications. Oxford University Press, Oxford (1987)
20. Ghindilis, A.L., Atanasov, P., Wilkins, E.: Enzyme-catalyzed direct electron transfer: fundamentals and analytical applications. Electroanalysis **9**, 661–674 (1997)
21. Willner, I.: Biomaterials for sensors, fuel cells, and circuitry. Science **298**, 2407–2408 (2002)
22. Marcus, R.A., Sutin, N.: Electron transfers in chemistry and biology. Biochem. Biophys. Acta. **811**, 265–322 (1985)
23. Eckermann, A.L., Feld, D.J., Shaw, J.A., Meade, T.J.: Electrochemistry of redox-active self-assembled monolayers. Coord. Chem. Rev. **254**, 1769–1802 (2010)
24. Carter, M.T., Rowe, G.K., Richardson, J.N., Tender, L.M., Terrill, R.H., Murray, R.W.: Distance dependence of the low-temperature electron-transfer kinetics of (ferrocenylcarboxy)-terminated alkanethiol monolayers. J. Am. Chem. Soc. **117**, 2896–2899 (1995)
25. Ramanavicius, A., Kausaite, A., Ramanaviciene, A.: Biofuel cell based on direct bioelectrocatalysis. Biosens. Bioelectron. **20**, 1962–1967 (2005)
26. Wilson, R., Turner, A.P.F.: Glucose oxidase: an ideal enzyme. Biosens. Bioelectron. **7**, 165–185 (1992)
27. Cai, C.X., Chen, J.: Direct electron transfer of glucose oxidase promoted by carbon nanotubes. Anal. Biochem. **332**, 75–83 (2004)
28. Wohlfahrt, G., Witt, S., Hendle, J., Schomburg, D., Kalisz, H.M., Hecht, H.-J.: 1.8 and 1.9 Å resolution structures of the *Penicillium amagasakiense* and *Aspergillus niger* glucose oxidase as a basis for modelling substrate complexes. Acta Crystallogr. Sect. D **55**, 969–977 (1999)
29. Katz, E., Buckmann, A.F., Willner, I.: Self-powered enzyme-based biosensors. J. Am. Chem. Soc. **123**, 10752–10753 (2001)
30. Tsujimura, S., Fujita, M., Tatsumi, H., Kano, K., Ikeda, T.: Bioelectrocatalysis-based dihydrogen/dioxygen fuel cell operating at physiological pH. Phys. Chem. Chem. Phys. **3**, 1331–1335 (2001)
31. Katz, E., Willner, I., Kotlyar, A.B.: A noncompartmentalized glucose/O_2 biofuel cell by bioengineered electrode surface. J. Electroanal. Chem. **479**, 64–68 (1999)
32. Karimian, N., Hashemi, P., Khanmohammadi, A., Afkhami, A., Bagheri, H.: The principles and recent applications of bioelectrocatalysis. Anal. Bioanal. Chem. Res. **7**, 281–301 (2020)
33. Degani, Y., Heller, A.: Electrical communication between redox centers of glucose oxidase and electrodes via electrostatically and covalently bound redox polymers. J. Am. Chem. Soc. **111**, 2357–2358 (1989)
34. Willner, I., Katz, E., Willner, B.: Electrical contact of redox enzyme layers associated with electrodes: Routes to amperometric biosensors. Electroanalysis **9**, 965–977 (1997)
35. Pereira, A.R., Luz, R.A.S., Dalmati, F.C.D.A., Crespilho, F.N.: Protein oligomerization based on Brønsted acid reaction. ACS Catal. **7**, 3082–3088 (2017)
36. Cracknell, J.A., Vincent, K.A., Armstrong, F.A.: Enzymes as working or inspirational electrocatalysts for fuel cells and electrolysis. Chem. Rev. **108**, 2439–2461 (2008)
37. Leech, D., Kavanagh, P., Schuhmann, W.: Enzymatic fuel cells: recent progress. Electrochim. Acta **84**, 223–234 (2012)
38. Clark, L.C., Lyons, C.: Electrode systems for continuous monitoring in cardiovascular surgery. Ann. N. Y. Acad. Sci. **102**, 29–000 (1962)

39. Rodrigues, V.C., Soares, J.C., Soares, A.C., Braz, D.C., Melendez, M.E., Ribas, L.C., Scabini, L.F.S., Bruno, O.M., Carvalho, A.L., Reis, R.M., Sanfelice, R.C., Oliveira, Jr., O.N.: Electrochemical and optical detection and machine learning applied to images of genosensors for diagnosis of prostate cancer with the biomarker PCA3. Talanta **222**, 121444 (2021)
40. Zeng, G.H., Xing, Y.B., Gao, J.A., Wang, Z.Q., Zhang, X.: Unconventional layer-by-layer assembly of graphene multilayer films for enzyme-based glucose and maltose biosensing. Langmuir **26**, 15022–15026 (2010)
41. Conzuelo, F., Gamella, M., Campuzano, S., Ruiz, M.A., Reviejo, A.J., Pingarron, J.M.: An integrated amperometric biosensor for the determination of lactose in Milk and dairy products. J. Agric. Food Chem. **58**, 7141–7148 (2010)
42. Khan, G.F., Wernet, W.: A highly sensitive amperometric creatinine sensor. Anal. Chim. Acta **351**, 151–158 (1997)
43. Kim, H.J., Bennetto, H.P., Halablab, M.A., Choi, C.H., Yoon, S.: Performance of an electrochemical sensor with different types of liposomal mediators for the detection of hemolytic bacteria. Sens. Actuat. B-Chem. **119**, 143–149 (2006)
44. Holland, J.T., Lau, C., Brozik, S., Atanassov, P., Banta, S.: Engineering of glucose oxidase for direct electron transfer via site-specific gold nanoparticle conjugation. J. Am. Chem. Soc. **133**, 19262–19265 (2011)
45. Frasconi, M., Heyman, A., Medalsy, I., Porath, D., Mazzei, F., Shoseyoy, O.: Wiring of redox enzymes on three dimensional self-assembled molecular scaffold. Langmuir **27**, 12606–12613 (2011)
46. Tee, K.L., Schwaneberg, U.: Directed evolution of oxygenases: screening systems, success stories and challenges. Comb. Chem. High Throughput Screen. **10**, 197–217 (2007)
47. Nazor, J., Schawaneberg, U.: Laboratory evolution of P450 BM-3 for mediated electron transfer. **4**, 638–644 (2006)
48. Tanaka, S., Igarashi, S., Ferri, S., Sode, K.: Increasing stability of water-soluble PQQ glucose dehydrogenase by increasing hydrophobic interaction at dimeric interface. BMC Biochem. **6**, 1 (2005)
49. Guven, G., Prodanovic, R., Schwaneberg, U.: Protein engineering—An option for enzymatic biofuel cell design. Electroanalysis **22**, 765 (2010)
50. Chen, L.Q., Zhang, X.E., Xie, W.H., Zhou, Y.F., Zhang, Z.P., Cass, A.E.G.: Genetic modification of glucose oxidase for improving performance of an amperometric glucose biosensor. Biosens. Bioelectron. **17**, 851–857 (2002)
51. Ali, M.H., Imperiali, B.: Protein oligomerization: how and why. Bioorg. Med. Chem. **13**, 5013–5020 (2005)
52. Nooren, I.M.A., Thornton, J.M.: Structural characterisation and functional significance of transient protein-protein interactions. J. Mol. Biol. 325 (2003)
53. Liu, Y., Eisenberg, D.: 3D domain swapping: as domains continue to swap. Protein Sci. **11**, 1285–1299 (2002)
54. Wang, W., Nema, S., Teagarden, D.: Protein aggregation—pathways and influencing factors. Int. J. Pharm. **390**, 89–99 (2010)
55. Mills, D.R., Peterson, R.L., Spielgeman, S.: An extracellular Darwinian experiment with a self-duplicating nucleic acid molecule. In: Proceedings of the National Academy of Sciences of the United States of America, vol. 58, p. 217 (1967)
56. Wong, T.S., Schwaneberg, U.: Protein engineering in bioelectrocatalysis. Curr. Opin. Biotechnol. **14**, 590–596 (2003)
57. Nazor, J., Dannenmann, S., Adjei, R.O., Fordjour, Y.B., Ghampson, I.T., Blanusa, M., Roccatano, D., Schewaneberg, U.: Laboratory evolution of P450 BM3 for mediated electron transfer yielding an activity-improved and reductase independent variant. Protein Eng. Des. Sel. **21**, 29–35 (2007)
58. Baik, S.H., Ide, T., Yoshida, H., Kagami, O., Harayama, S.: Significantly enhanced stability of glucose dehydrogenase by directed evolution. Appl. Microbiol. Biotechnol. **61**, 329–335 (2003)

59. Boublik, Y., Saint-Aguet, P., Lougarre, A., Arnaud, M., Villatte, F., Estrada-Mondaca, S., Fournier, D.: Acetylcholinesterase engineering for detection of insecticide residues. Protein Eng. Des. Sel. **15**, 43–50 (2002)
60. Sotiropouulou, S., Fournier, D., Chaniotakis, N.A.: Genetically engineered acetylcholinesterase-based biosensorfor attomolar detection of dichlorvos. Biosens. Bioelectron. **20**, 2347–2352 (2005)
61. Minteer, S.D., Liaw, B.Y., Cooney, M.J.: Enzyme-based biofuel cells. Curr. Opin. Biotechnol. **18**, 228–234 (2007)
62. Heller, A.: Electron-conducting redox hydrogels: design, characteristics and synthesis. Curr. Opin. Chem. Biol. **10**, 664–672 (2006)
63. Welch, C.M., Compton, R.G.: The use of nanoparticles in electroanalysis: a review. Anal. Bioanal. Chem. **384**, 601–619 (2006)
64. Habermuller, L., Mosbach, M., Schuhmann, W.: Electron-transfer mechanisms in amperometric biosensors. Fresenius J. Anal. Chem. **366**, 560–568 (2000)
65. Willner, I., Willner, B.: Biomaterials integrated with electronic elements: en route to bioelectronics. Trends Biotechnol. **19**, 222–230 (2001)
66. Ivnitski, D., Branch, B., Atanassov, P., Apblett, C.: Glucose oxidase anode for biofuel cell based on direct electron transfer. Electrochem. Commun. **8**, 1204–1210 (2006)
67. Gilardi, G., Fantuzzi, A., Sadeghi, S.J.: Engineering and design in the bioelectrochemistry of metalloproteins. Curr. Opin. Struct. Biol. **11**, 491–499 (2001)

Electroenzymatic Redox Organic Synthesis

Roberto da Silva Gomes

Abstract Electroenzymatic processes combine interface enzymes and electrodes to reduce or oxidize molecules. The technique involving electroenzymatic technology can be applied to solve a variety of problems by producing biosensors, biocapacitors, and synthesizing molecules with biological interest such as chiral molecules, renewable energy, and studying reaction mechanisms. This process involves coupling an enzyme on an electrode where electrons are transferred between the electrode and the enzyme cofactor. Enzymatic electrosynthesis has attracted significant attention due to the possibility of its use as renewable electrical energy that can drive oxidative and/or reductive reactions by enzymes. Along with the mild condition where reactions occur, its high selectivity is also desirable. Specific products can be achieved at low cost, replacing the lengthy and costly processes employed in traditional chemical reactions. In this chapter, we will discuss some relevant redox enzymes and methodologies involved in the formation of the enzyme-electrode and their role in the electron transfer for the oxidation/reduction of C, O, N, and H atoms.

1 Introduction

Bioelectrochemical synthesis comprises different techniques and approaches that interface with enzymes and electrodes [10]. This enzyme-electrode system has attracted the attention of researchers due to its versatile applicability, such as biosensors [10], energy conversion and storage (such as biofuel cells) [57, 87], and biocapacitors [57, 72]. Moreover, the enzyme-electrode system can be used in electrochemical reaction systems to provide the necessary redox environment equivalent to enzymatic transformations. This enzyme-electrode biotransformation that regenerates a cofactor or cosubstrate by an electrochemical reaction system is called electroenzymatic synthesis [37, 76, 77]. The concept of this electrochemical system is to employ an enzyme-electrode in such a way that this electrode can substitute one

R. da Silva Gomes (✉)
Department of Pharmaceutical Sciences, North Dakota State University, Fargo, ND 58105, USA
e-mail: roberto.gomes@ndsu.edu

F. N. Crespilho (ed.), *Advances in Bioelectrochemistry Volume 4*,
https://doi.org/10.1007/978-3-030-99662-8_2

of an enzyme's substrate, transferring electrons to or from to support the oxidation or reduction of a second substrate [10, 37, 72, 76, 77, 87].

Another advantage in employing enzymes is because they show remarkable chemo-, regio-, and stereoselectivity and catalyze organic syntheses without significant side reactions [84]. Additionally, biocatalytic redox reactions offer selectivity and yields achievable with conventional chemical synthesis. For example, the use of the broad oxidoreductases covers dehydrogenation, oxygenation disproportionation, oxidative bond formation, and single-electron transfer [84].

The Redox reaction of nitrogen gas (N_2), carbon dioxide (CO_2), hydronium (H^+), carboxylates such as formate ($HCOO^-$), and hydrogen gas (H_2) represents a very important research field, such as atom-economy in the reaction system and clean energy [68]. In this chapter, we will discuss some relevant redox enzymes and methodologies involved in the formation of the enzyme-electrode and their role in the electron transfer for the oxidation/reduction of C, O, N, and H atoms.

2 Oxidase Enzymes

Oxidases are a group of enzymes capable of catalyzing the oxidation of C–N and C–O bonds, especially those involving dioxygen (O_2) as an electron acceptor. They are mainly flavin-dependent enzymes catalyzing oxidation reactions. They use flavin adenine dinucleotide (FAD) as a cofactor. This cofactor does not allow diffusion to the electrode once they are strongly bound to the enzyme. Additionally, FAD can be regenerated by reducing oxygen to hydrogen peroxide (H_2O_2) [s]. In this reaction, the H_2O_2 formation as byproduct results in oxidase stability problems that decrease the effectiveness of the method. To overcome this problem, a catalase is used to compete with the excess of formed H_2O_2 [90]. Another approach to avoid the problem with H_2O_2 formation to regenerate the cofactor is to react under anaerobic conditions. Electron transfer to FAD is necessary once the FAD biding site and the location of the active site do not allow direct electron transfer, which requires a mediator (such as ferrocene) to overcome this problem [4].

The first report described by Hill et al. is the oxidation of p-cresol **1** into *p*-hydroxybenzaldehyde **2** with *p*-cresol methylhydroxylase (PCMH) using *Pseudomonas alcaligenes* (Scheme 1) [4, 23].

In this case, the mediator was a combination between ferrocene boronic acid and a gold electrode, which converted 85%. The same protocol was used but with PEG2000 ferrocene and graphite electrode as a mediator with 84% overall conversion [19]. Both ferrocene boronic acid and PEG2000 ferrocene were used as a sort of "stationary phase" held in the reaction chamber, while product and substrate pass through the system to provide a continuous process. The authors used an electrochemical enzyme membrane reactor (EEMR). In this reactor, an ultrafiltration membrane is attached to an electrochemical cell with carbon felt that works to guarantee the retention of both mediator and enzyme in this flow system. The final product **2** could be controlled by controlling the residence time. At residence time of 3 **h**, 95% of conversion

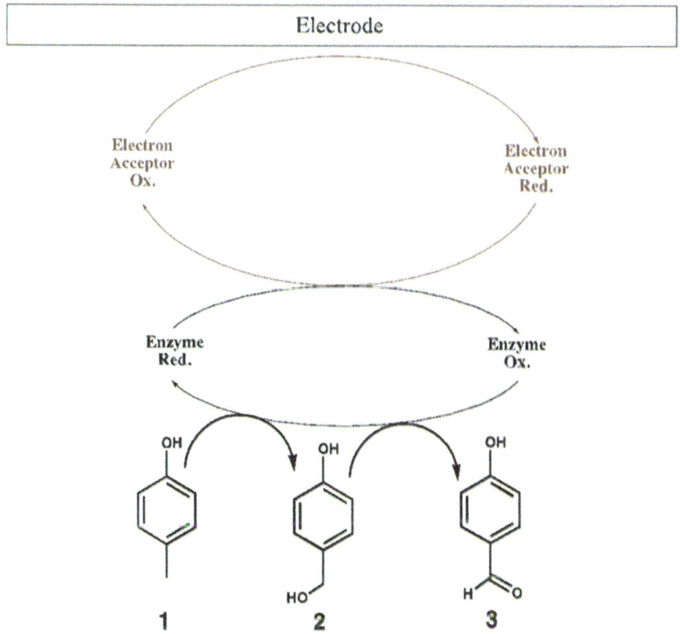

Scheme 1 Bioelectrotransformation of p-cresol into p-hydroxybenzaldehyde

was observed, while intermediate **3** was the main product of this reaction at lower residence time.

Brielbeck et al. used the same PCMH for the oxidation of 4-ethylphenol **4** to provide the chiral intermediate (*S*)-4-(1-hydroxyethyl)-phenol **5**, achieving 70% conversion and 88% *ee* (Scheme 2a). In the same system, the use of p-ethylphenol methylene hydroxylase (EPMH) provided the enantiomer *R* **6** with a 100% conversion rate (Scheme 2b) [2].

The use of ionic liquids was explored by Kohlmann et al. to improve the electroenzymatic synthesis processes [35]. In addition to increasing the conductivity of the reaction and stabilizing the catalyst, ionic liquids can enhance the solubility of

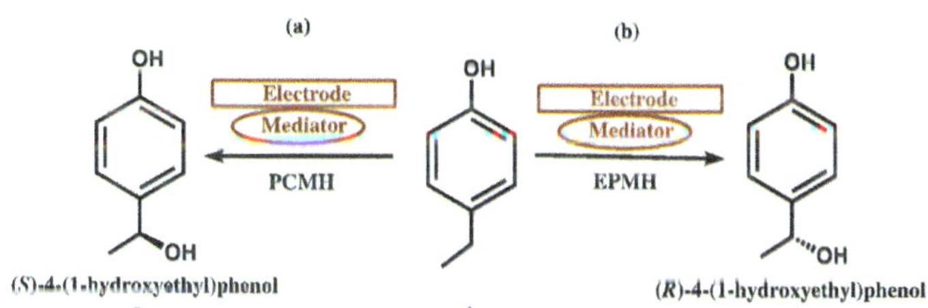

Scheme 2 **a** Conversion of 4-ethylphenol in (*S*)-4-(1-hydroxyethyl)-phenol using PCMH and **b** Conversion of 4-ethylphenol in (*S*)-4-(1-hydroxyethyl)-phenol using EPMH

substrates, a common problem in these systems. Kohlmann et al. reported the use of ionic liquid in the resolution of *D,L*-methionine catalyzes by *D*-amino acid oxidase. The use of the ionic liquid [MMIM] [Me_2PO_4] provided an optical purity of > 99% *ee*.

Usually, FAD-dependent amino acid oxidase (FAD-AAO) is used to produce amino acids by reduction of ketones. FAD-AAO oxidizes amino acids to imino acids which are rapidly hydrolyzed to α-keto acids. However, Kawataba et al. reported the use of a redox potential more negative of FAD along with 1-aminopropyl-1′-methyl-4,4′-dipyridinium iodide (ADPy) could revert the oxidation of amino acids. In this case, the use of a glassy carbon electrode to immobilize the mediator could perform the reduction of pyruvic acid and phenyl pyruvic acid into *D*-alanine and *L*-phenylalanine, respectively [32].

One direct electrochemical regeneration of FAD by using an electrode without a mediator was described by Jeon et al. [31]. The reaction system was set up in a two-chamber electrochemical reactor with graphite-Cu (II) as anode and graphite-Fe (III) electrode as a cathode to convert xylitol to xylose using *Escherichia Coli* and *Saccharomyces cerevisiae* as biocatalysts.

3 Monooxygenases

Monooxygenases can transfer one atom of molecular oxygen to X–H bonds when the other atom is reduced to water. They are also involved in the catalysis of several reactions, such as hydroxylations, epoxidations, halogenations, and Baeyer–Villiger oxidations [9, 14, 75]. Usually, the oxidation of NADPH provides the electrons required for the reduction step. As for the regeneration of NADPH-dependent enzymes, several mediators are reported in the literature [68], such as cobalt(III) sepulchratetrichloride [13, 15], and 1,1′-dicarboxy-cobaltocene [81].

The first attempt to use monooxygenase in bioelectrochemical synthesis was reported by Faulkner et al. The use of a purified protein rFP450(mRat4Al/mRat-OR) L1, a P450 monooxygenase fused with a rat NADPH-P450 reductase, provided the required two electrons for reductive oxygen activation from an organometallic mediator (CoSep) instead of the natural NADPH for the ω-hydroxylation of lauric acid [15]. Subsequently, Estabrook et al. reported the use of the same system but replacing NADPH-P450 reductase by cytochrome P450 monooxygenase [13].

O-hydroxylation of several α-substituted phenol derivatives was achieved using 2-hydroxybiphenyl-3-monooxygenase (HbpA). In this case, the electrochemical regeneration of NADH was possible due to the attachment of the enzyme to [Cp*Rh(bpy)Cl]Cl complex that was used as a mediator. However, the usage of [Cp*Rh(bpy)Cl]Cl allowed the formation of hydrogen peroxide. Molecular oxygen reacts with the rhodium complex or even can be reduced at the cathode [24].

Direct electrochemical regeneration of NADPH without electron transfer mediator to FAD was performed in the chiral (*S*)-epoxidation of styrenes by the flavin-dependent styrene monooxygenase (StyA) from *Pseudomonas sp.* VLB120 [25]. In this protocol, the system containing three enzymes and the two cofactors could be reduced to StyA and FAD.

Ruinatscha et al. designed a reactor separating the high concentration of molecular oxygen from the cathode chamber to increase cofactor regeneration rates [64]. This flow reaction reactor provided electrodes with a maximal volumetric surface that could achieve 0.3–1.3 mM h^{-1} regeneration of FAD in the StyA system. Despite all efforts to avoid oxygen reduction, this oxigen transformation is still a problem when electrochemical synthesis is performed with enzymes dependent on molecular oxygen. Even the use of the more oxygen stable $[Cp^*Rh(I)(bpy)H]^+$ by Tosstorff et al. for the conversion of *p*-xylene to 2,5-dimethylphenol by P450 BM3 monooxygenase and cobalt sepulchrate as mediator showing oxygen depletion which appeared to solve the problem with the oxygen reduction, the system presented a loss in energy efficiency [80].

Several mediators are reported to transfer electrons artificially by cofactor generation with monooxygenases. Aiming for a more efficient enzymatic process, Strohle et al. generated a virtual protocol to dock mediator candidates for P450cin [78]. The protocol was able to predict suitable mediators binding affinity. The selected mediators provided 70% fitting with the prediction studies. Along with the virtual prediction, several other parameters must be evaluated, such as applied potential or concentrations of mediator and enzyme. Their optimization could lead to higher product formation yield. Ley et al. developed the electrochemical microtiter plate (eMTP). This methodology can elucidate the influence of P450 BM3 and mediator concentrations and apply a specific potential according to the experiment's needs [44].

4 Peroxidases

Unlike monooxygenases, peroxidases are not dependent on cofactors such as NADPH or FADH. Peroxidases contain a porphyrin group (heme), vanadate, or manganese as chelating metal. Usually, peroxidases use hydrogen peroxide as an electron acceptor to catalyze substrate oxidation. As discussed previously, the use of hydrogen peroxide as an electron acceptor decreases the stability of the enzyme if it is exposed to high concentrations. Fortunately, several approaches are reported to prevent high local concentrations of H_2O_2, such as electrochemical generation of hydrogen peroxide [59], in situ synthesis of hydrogen peroxide in combination with a second enzyme [82], and sensor-controlled feeding of hydrogen peroxide [70]. The addition of a controlled H_2O_2 concentration to the enzymatic system by varying the electrochemical parameters makes the electrochemical generation of hydrogen peroxide the most common method [59]. Here, no mediator is required, and the cathodic reduction of oxygen forms H_2O_2, making the system simple. Mercury, gold, and carbon can be used as electrodes. Because it is safe and economically attractive, carbon is the most common material for bioelectrochemical synthesis [27, 28, 39].

The first example is reported by Chen et al. [7]. Free horseradish peroxidase (HRP) promoted the *N*-demethylation of *N*,*N*-dimethylaniline (DMA). A high concentration of H_2O_2 and formaldehyde was observed by increasing the applied current. Lee et al., using another approach, oxidized veratryl alcohol to veratraldehyde with lignin peroxidase [43]. In this reaction, the cathode and the anode are separated by a cation-exchange membrane in a two-chamber reactor. A comparison between the rates of veratraldehyde formation with electrochemically hydrogen peroxide generation and the natural hydrogen peroxide method was performed. The natural H_2O_2 generation method resulted in a higher veratraldehyde formation rate than the electrochemical method.

Chlorination of barbituric acid to 5-chlorobarbituric acid, among other reactions, can be achieved using heme-dependent chloroperoxidase (CPO) from the filamentous fungus *Caldariomyces fumago* [42]. A bioelectrolytic system containing an electrolytic cell for H_2O_2 generation, a hollow-fiber membrane for biocatalysis, and an anion exchanger for product collection were used in the first attempt. After 24 h, 0.33 mM of 5-chlorobarbituric acid was produced with an initial reaction rate of 0.17 mM h^{-1}.

Dembitsky et al. reported that CPO is also used to mediate the formation of sulfoxides from the oxidation of dialkyl sulfides [11]. In a chiral point of view for the formation of sulfoxides, Lutz et al. reported an in situ electrogeneration for the chiral oxidation of thioanisole to (*R*)- methylphenyl sulfoxide with high optical purity [46]. To increase the yield of the chiral (*R*-methylphenyl sulfoxide, they used a three-chamber cell with a flow-by-electrode system and two anodes [47]. In this optimized reaction system, the reaction rate could be comparable to sensor-controlled dosing of hydrogen peroxide. The use of 2% of [EMIM][$EtSO_4$] for the oxidation of thioanisole to (*R*)-methylphenyl sulfoxide increased by 4.2 the productivity [91].

The success of the formations of (*R*)-methylphenylsulfoxide led Lutz *et al.* [36] to report an extension of the synthesis of sulfoxides, such as *N*-MOC(methoxycarbonyl)-*L*-methionine methyl ester sulfoxide, (*R*)-methoxyphenylmethyl sulfoxide, and (*R*)-methyl-*p*-tolylsulfoxide. The reactions resulted in a conversion of 60% of *N*-MOC-*L*-methioninemethyl ester sulfoxide, 83% of (*R*)-methoxyphenylmethyl sulfoxide, and 76% of (*R*)-methyl *p*-tolylsulfoxide, respectively [26] (Scheme 3).

H_2O_2 generation is directly impacted by limitation in mass transport due to a limited electrode surface. A gas diffusion electrode (GDE) was reported by Krieg et al. as a possibility to overcome this H_2O_2 generation. In this case, the oxygen can be reduced to H_2O_2 by GDEs by using a solid, liquid, and gaseous interface that increases the problem with the limitation of the electrode surface [39]. The CPO-catalyzed monochlorodimedone chlorination and thioanisole and indole oxidation in a GDE-based system resulted in a *ttn* and STY improvement [39]. Later, in another study, Holtmann et al. used a GDE system to chlorinate monochlorodimedone (MCD) [27]. Here, using the GDE-based reaction system for H_2O_2-dependent enzymes led to high STY and *ttn*. Finally, the heme-thiolate peroxygenase AaeUPO from the fungus *Agrocybe aegerita* can catalyze the hydroxylation of ethyl-benzene into 1-phenethyl alcohol [28]. This GDE-based system using an electrochemical H_2O_2 electrode led

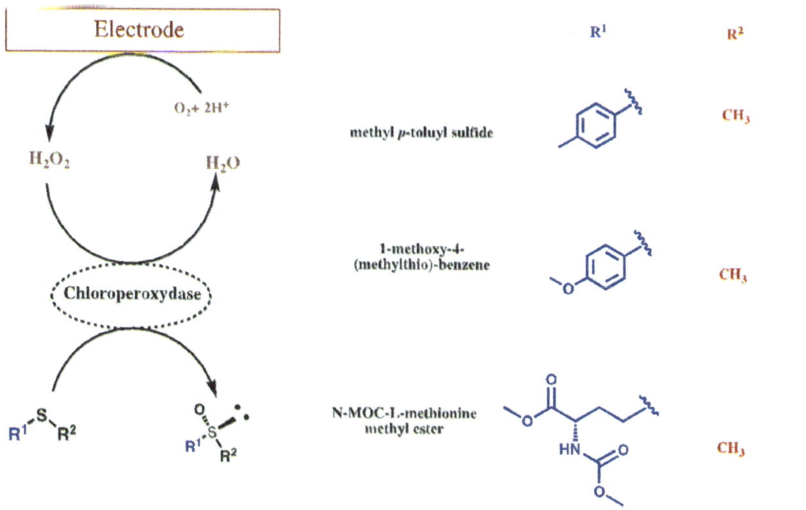

Scheme 3 Hollmann Conversion of regular sulfides into chiral sulfoxides

to a high reaction rate which can be compared with the productivities of AaeUPO with a non-electrochemical hydrogen peroxide source.

5 Nitrogenase

Nitrogenase has gained interest due to its ability to reduce N_2 to ammonia (NH_3) in bioelectrochemical synthesis at ambient temperature and pressure in the opposite way of the Haber–Bosch process that uses high-temperature pressure [84]. It is estimated that around 50% found in the human body originates from Haber–Bosch [12]. This process uses a Fe-based catalytic system to break the triple bond of the inert N_2 at high temperature and pressure, including the production of H_2, to make this process highly efficient. Additionally, it is estimated that this process consumes 1–2% of the global energy source producing 3% of all CO_2 worldwide [17, 34]. Conversely, biological N_2 reduction can be performed by a metal–enzyme complex found in select bacteria and archaeas named nitrogenase. This metal–enzyme complex works in an electron-transfer system where adenosine triphosphate (ATP)-hydrolyzing iron protein (Fe protein) and a catalytic molybdenum-containing protein (MoFe protein) can reduce or "fix" N_2.

Significant advances in electron transfer from electrodes to MoFe/VFe/FeFe metalloenzymes for electrocatalytic reduction of different substrates, such as H^+, N_2, CO, cyanide (CN^-), azide (N_3), and nitrite (NO_2), highlight MoFe/VFe/FeFe metalloenzymesas strong candidates for electroenzymatic synthesis [50].

Minteer and Seefeldt used the previously described low-potential mediator cobaltocene to transfer electrons between an electrode and MoFe protein [52]. Using $E^0 =$

−96 V, the mediator was able to provide the required electrons to support the reduction of hydronium (H^+)–H_2 and nitrogen reduction of N_3^- and NO_2^-–NH_3. The same mediator was adapted to provide electrons to reduce H^+ and CO_2 by using MoFe- or FeFe-enzyme complex achieving H_2 and $HCOO^-$, respectively [29]. The formation of ethene and propene by the formation of C–C bond in a cobaltocene-mediated electrosynthetic system using VFe-enzyme complex by reduction of CO_2 also was explored. It is also reported that Fe/VFe protein coupling and ATP hydrolysis can catalyze the C–C bond from CO reduction [30].

Lastly, the cobaltocene mediator-based methodology was used to study the rate-limiting step of the MoFe protein in the catalytic reduction of H^+. The results provided an insight that the rate-limiting step is related to the release of inorganic phosphate of the Fe-protein complex [69, 85]. Pyrene functionalized polymer in a DET/MoFe protein system was developed by Hickey et al., aiming to reduce N_2 independent of Fe-protein or ATP-hydrolysis, which opened a new gate of possibilities for the reduction of N_2 by nitrogenase [21, 22].

In a search for an electron delivery to the MoFe protein, Milton et al. proposed an electrochemical methodology that delivers electrons to the Fe protein, which also provides electrons to the MoFe with the hydrolysis of ATP [51]. In this approach, between electrodes and Fe protein was adapted MV as an electron mediator that could transfer the required number of electrons to the Fe protein showing a promising method to reduce N_2 in a renewable energy system. At 0.61 $_V$, Rhode and Lukoyanov reported a reduction of N_2 at a Faradaic efficiency (FE) of 59%, which means that the nitrogenase used 59% of the available electrons to convert N_2 in NH_3 [44, 63, 65].

A 5-enzyme cascade reaction system can also be used [5]. In this system, Fe protein, MoFe protein, diaphorase, *L*-alanine dehydrogenase, and ω-transaminase reduce N_2 to NH_3. It was reported by Minteer et al. that the 5-enzyme cascade system using low MV and nitrogenase successfully formed the chiral (*R*)-1-methyl-3-phenylpropylamine. This idea was adapted to an ATP hydrolysis-dependent system containing H_2ase-based H_2-oxidizing anode coupled to a nitrogenase-based cathode to reduce N_2 to NH_3. The same method was adapted again with diaphorase and leucine dehydrogenase to provide chiral amino acids from α-ketoacids in good yield and *ee* [6].

6 CO_2-Reductase

CO_2 atmospheric accumulation plays a crucial role in world climate change and warming. The generation of CO_2 by fossil fuel combustion (among other human and industrial sources) keeps increasing. Currently, it is common sense that something has to be done to decrease CO_2 levels to maintain a livable condition on the surface of the earth. CO_2 reduction by an enzymatic electrochemical process is an environmental friend possibility. In this process, an enzyme catalyzes CO_2 reduction by electron transfer systems. Many enzymes can be found in nature, especially in microorganisms and plants. Due to their ability to catalyze the reduction of CO_2 to CO, they are commonly termed carbon dioxide reductases [67]. Formate dehydrogenases (Fdh) have been highlighted for their ability to reduce CO_2 to formate ($HCOO^-$) [61]. Fdh is easily found in nature within two categories: (a) metal-independent (MI) and metal-dependent (MD) Fdhs [61].

Interestingly, because MI-Fdhs are air-stable, they do not suffer inhibition by O_2 and can be used in a wide range of pH conditions without losing their activity. A negative point of MI-Fdhs is when compared with MD-Fdhs, they present a weaker activity, and consequently, they tend to oxidize $HCOO^-$ to CO_2 instead of the contrary. Additionally, their electrochemical immobilization is hampered by their NADPH dependency. Another problem is the formation of the inactive NAD_2 dimer that is a consequence of the electroreduction of NAD^+ [24]. MD-Fdhs contain a considerable amount of iron-sulfur (FeS) clusters responsible for transferring electrons to the active site, which usually includes two divalent pyranopterin cofactors coordinated to a Mo or W atom, a selenocysteine or cysteine residue with a site occupied by a sulfur atom. Due to exposed above, MD-Fdh's tend better to CO_2 reduction than MI-Fdhs [8].

If the enzymatic immobilization leads to correct orientation, DET can be observed with MD-Fdhs. As mentioned before, MD-Fdhs are good candidates for CO_2 reduction with high FEs, especially because they are not expected to catalyze H^+ reduction. In 2008, Reda et al. reported the use of MD-Fdh DET to electroenzymatically reduce CO_2 attached to a W-dependent Fdh from *Desulfovibrio gigas* immobilized to an electrode [60]. Interestingly, the results showed FEs of around 100%, and the enzyme could keep its catalytic activity at a pH ranging from 4 to 8. An extension of the enzyme scope was reported to explain the inhibition mechanism of the MD-Fdh from *Escherichia coli* with different inhibitors in the oxidation of formate or reduction of CO_2 by DET [62]. Another study proposed a study of DET when the system used a W-dependent formate dehydrogenase (FoDH1) from *Methylobacterium extorquens* immobilized in gold nanoparticles (AuNPs) coated with mercaptopyridine moieties and Ketjen Black carbon particles for formate oxidation and CO_2 reduction. Because these enzymes possess a non-covalently bound flavin mononucleotide (FMN), this protocol favored the oxidation and reduction of NAD/NADH [62, 66]. Another study of DET used a W-dependent Fdh from *Clostridium ljungdahlii* embedded in a conductive polyaniline (PANI) hydrogel. The use of different PANI hydrogel structures allowed the authors to infer that the residues possibly impacted the electron transfer from the hydrogel-coated electrode to the W active site [40].

As discussed previously in this chapter, nitrogenases can also reduce CO_2 into HCOO in a cobaltocene-mediated system [29]. Cobaltocene can be attached to a poly(allylamine) chain providing a $E^{0'} = 0.58$ V redox polymer termed Cc-PAA. The attachment of Cc-PAA onto Mo-dependent Fdh from *E. coli* showed an FE of 99% for CO_2 reduction to $HCOO^-$ [88]

Parkinson et al. reported the MV-reduction by photo-excitation of electrons from a p-type indium phosphide photo-electrode that consequently reduced CO_2 to $HCOO^-$ by *Pseudomonas oxalaticus* Fdh [58]. The same idea was further used by Amao et al. to describe the efficient use of an ITO electrode connected to a long alkyl chain that contained MV embedded with *Saccharomyces cerevisiae* Fdh to reduce CO_2 to produce $HCOO^-$ that is size chain-dependent [1].

Improvements were made to increase the reduction of CO_2 to methanol (CH_3OH). Some of these methodologies are described below. In this process, the first step is usually the reduction of CO_2 by Fdh to provide $HCOO^-$. The following step uses formaldehyde dehydrogenase to catalyze the reduction of $HCOO^-$ to formaldehyde (CH_2O), and the last step involves alcohol dehydrogenase to catalyze CH_2O reduction CH_3OH [41, 56].

7 H_2ase

Because it is renewable and clean, H_2 is considered a promising alternative energy source to replace fossil fuels once only water is the side product. It is well known that its use is minimal due to issues with storage and transportation. Its high energetic bond (dissociation energy $\Delta H = +436$ kJ mol^{-1}) increases its explosive potential when mixed with O_2, for example. Another problem is using the expensive H^+ and rare noble metal electrodes in H_2 cell fuel. Due to these facts, a lot of effort has been used on systems that can produce H_2 under mild conditions to replace fossil fuels [3].

Many microorganisms have some enzymatic pathways to provide and oxidize H_2 at low potentials using mainly a metalloenzyme termed H_2ase enzymes that typically reversibly catalyze the interconversion of H_2 and $2H^+$. In the field of enzymatic electrochemical systems to produce H_2, the most studied H_2ase enzymes are [FeFe]- and [NiFe]–H_2ases which are conventionally named according to the nature of metallic cofactor in the active site. Because they possess FeS clusters, they can transport electrons through the protein between the active sites [16]. [FeFe]-H_2ases are typically more suitable for H^+ reduction at electrodes due to their low affinity for H_2 [18] when compared to [NiFe]–H_2ases that favor H_2 oxidation. H_2ases in enzymatic electrochemistry are mostly reported for anodic enzymatic electrochemistry, where H_2 is oxidized as a fuel to provide a source of energy [51, 73, 74]. Additionally, H_2ases have been intensely studied by electrochemistry in the last three decades [83, 86].

After enzymatic adsorption, the catalytic activity can be lost due to structural changes in the electroenzymatic system on the surface of the electrode due to the interfacial interactions which favors the maintenance of the stability of the enzyme by immobilization important for the electrochemical system. The use of immobilized H_2ases was reported Lojou et al. [48, 53]. The O_2 resistant [NiFe]-H_2ases from the hyperthermophilic bacteria *Aquifex aeolicus* (Aa) were immobilized on various electrode surfaces.

[NiFe]–H_2ases are known to support H2 oxidation better and usually are inactivated in the presence of O_2. Goldet et al. showed that *Ralstonia* [NiFe]–H_2ases could catalyze H_2 production in the presence of O_2 and in the air itself, as long as H_2 is continuously removed [20]. Some DET usage also has been explored using single-walled carbon nanotubes for the H_2ase immobilization [33, 49, 79]. King's group reported a current density for H_2 formation around 2 mA cm^2 with a FE of 98% using DET to immobilize [FeFe]-H_2ase from *Clostridium perfringens* by simple adsorption on anatase TiO_2 electrodes without losing activity for at least a week [54].

For MET, in electroenzymatic H_2 formation: [FeFe]–H_2ase from the photosynthetic green alga *Chlamydomonas reinhardtii* was immobilized on a gold electrode coated with 3-mercaptopropionic acid or 11-mercapto propionic acid SAMs. In this case, the length of the carbon chain played an important role. The size of the carbon chain drives the electrostatic interaction and the flexibility of the SAM layer leading to several different catalytic activities depending on what SAMs are employed, primarily due to the changes in the surface structure and orientation of the H_2ase [38].

In another approach, Oh et al. reported a two-chamber fuel cell with H_2ase from *Thiocapsa roseopersicina* immobilized at the cathode for H^+ reduction. In the anodic chamber, dithionite-based electron source and MV was employed as an electron relay for both the anodic and cathodic chambers showing a low rate for the production of H_2 when compared to the production of H_2 by H_2ase of this strain [55].

In 2018, Shiraiwa et al. reported another template for polymeric immobilization [71]. MV-based redox polymer and [NiFe]–H_2ase from *D. vulgaris* Miyazaki F were co-immobilized on glassy carbon, and carbon felt working electrodes. In the resulting system, the high potential reductive reactivation of the Ni–A/Ni–B states was achieved by reduction with the MV moieties on the redox polymer and the oxidative inactivation of states. High current densities H^+ reduction was observed for the same electrochemical system by cyclic voltammetry using larger carbon felt electrodes, promising to be a new open gate for exploring the preparation of H_2 in electroenzymatic systems [71].

References

1. Amao, Y., Shuto, N.: Formate dehydrogenase-viologen-immobilized electrode for CO_2 conversion to develop an artificial photosynthesis system. Res. Chem. Intermed. **40**, 3267–3276 (2014)
2. Brielbeck, B., Frede, M., Steckhan, E.: Continuous electroenzymatic synthesis employing the electrochemical enzyme membrane reactor. Biocatalysis **10**, 49–64 (1994)
3. Cadoux, C., Milton, R.: Recent enzymatic electrochemistry for reductive reactions. Chem. Electrochem. **7**, 1974–1986 (2020)
4. Cass, A.E., Davis, G., et al.: Ferrocene-mediated enzyme electrode for amperometric determination of glucose. Anal. Chem. **56**, 667–671 (1984)
5. Chen, H., Cai, R., et al.: Upgraded bioelectrocatalytic N_2 fixation: from N_2 to Chiral Amine intermediates. J. Am. Chem. Soc. **141**, 4963–4971 (2019)
6. Chen, H., Prater, M.B., et al.: Bioelectrocatalytic conversion from N_2 to Chiral Amino acids in a H_2/α-Keto acid enzymatic fuel cell. J. Am. Chem. Soc. J. Am. Chem. Soc. **142**, 4028–4036 (2020)
7. Chen, J.K., Nobe, K.: Oxidation of dimethylaniline by horseradish-peroxidase and electrogenerated peroxide. J. Electrochem. Soc. **140**, 299–303 (1993)

8. Choe, H., Joo, J.C.: Efficient CO_2-reducing activity of NAD-dependent formate dehydrogenase from Thiobacillus sp. KNK65MA for formate production from CO_2 Gas. PLoS One **9**, e103111 (2014)
9. Cirino, P.C., Arnold, F.H.: Protein engineering of oxygenases for biocatalysis. Curr. Opin. Chem. Biol. **6**, 130–135 (2002)
10. Das, P., Das, M., et al.: Recent advances on developing 3rd generation enzyme electrode for biosensor applications. Biosens. Bioelectron. **79**, 386–397 (2016)
11. Dembitsky, V.M.: Oxidation, epoxidation and sulfoxidation reactions catalysed by haloperoxidases. Tetrahedron **59**, 4701–4720 (2003)
12. Erisman, J.W., Sutton, M.A., et al.: How a century of ammonia synthesis changed the world. Nat. Geosci. **1**, 636–639 (2008)
13. Estabrook, R.W., Faulkner, K.M., et al.: Application of electrochemistry for P450-catalyzed reactions. Methods Enzymol. **272**, 44–51 (1996)
14. Faber, K.: Biotransformations in Organic Chemistry, 5th edn. Springer, New York (2000)
15. Faulkner, K.M., Shet, M.S., et al.: Electrocatalytically driven omega-hydroxylation of fatty-acids using cytochrome-P450 4A1. PNAS **92**, 7705–7709 (1995)
16. Fontecilla-Camps, J.C., Volbeda, A., et al.: Structure/function relationships of [NiFe]and [FeFe]-hydrogenases. Chem. Rev. **107**, 4273–4303 (2007)
17. Foster, S.L., Bakovic, S.I.P., et al.: Catalysts for nitrogen reduction to ammonia. Nat. Can. **1**, 490–500 (2018)
18. Fourmond, V., Baffert, C.: The mechanism of inhibition by H_2 of H_2-evolution by hydrogenases. Chem. Commun. **49**, 6840–6842 (2013)
19. Frede, M., Steckhan, E.: Continuous electrochemical activation of flavoenzymes using polyethyleneglycol-bound ferrocenes as mediators—A model for the application of oxidoreductases as oxidation catalysts in organic synthesis. Tetrahedron Lett. **32**, 5063–5066 (1991)
20. Goldet, G., Wait, A.F., et al.: Hydrogen production under aerobic conditions by membrane-bound hydrogenases from ralstonia species. J. Am. Chem. Soc. **130**, 11106–11113 (2008)
21. Hickey, D.P., Cai, R.: Establishing a thermodynamic landscape for the active site of Mo-dependent nitrogenase. J. Am. Chem. Soc. **141**, 17150–17157 (2019)
22. Hickey, D.P., Lim, K., et al.: Pyrene hydrogel for promoting direct bioelectrochemistry: ATP-independent electroenzymatic reduction of N_2. Chem. Sci. **9**, 5172–5177 (2018)
23. Hill, H.A.O., Oliver, B.N., et al.: The enzyme-catalyzed electrochemical conversion of para-cresol into para-hydroxybenzaldehyde. J. Chem. Soc. Chem. Commun. **21**, 1469–1471 (1985)
24. Hollmann, F., Arends, I.W.C.E., et al.: Biocatalytic redox reactions for organic synthesis: nonconventional regeneration methods. Chem. Cat. Chem. **2**, 762–782 (2010)
25. Hollmann, F., Hofstetter, K., et al.: Direct electrochemical regeneration of monooxygenase subunits for biocatalytic asymmetric epoxidation. J. Am. Chem. Soc. **127**, 6540–6541 (2005)
26. Hollmann, F., Schmid, A., Steckhan, E.: The first synthetic application of a monooxygenase employing indirect electrochemical NADH regeneration. Angew. Chem. Int. Ed. **40**, 169–171 (2001)
27. Holtmann, D., Krieg, T., et al.: Electroenzymatic process to overcome enzyme instabilities. Catal. Commun. **51**, 82–85 (2014)
28. Horst, A.E.W., Bormann, S., et al.: Electro-enzymatic hydroxylation of ethylbenzene by the evolved unspecific peroxygenase of Agrocybe aegerita. J. Mol. Catal. B Enzym. **133**, S137–S142 (2016)
29. Hu, B., Harris, D.F., et al.: Electrocatalytic CO_2 reduction catalyzed by nitrogenase MoFe and FeFe proteins. Bioelectrochemistry **120**, 104–109 (2018)
30. Hu, Y., Lee, C.C., et al.: Extending the carbon chain: hydrocarbon formation catalyzed by vanadium/molybdenum nitrogenases. Science **333**, 753–755 (2011)
31. Jeon, J.S., Shin, I.H., et al.: Electrochemical regeneration of FAD by catalytic electrode without electron mediator and biochemical reducing power. J. Microbiol. Biotechnol. **15**, 281–286 (2005)

32. Kawabata, S., Iwata, N., Yoneyama, H.: Asymmetric electrosynthesis of amino acid using an electrode modified with amino acid oxidase and electron mediator. Chem. Lett. **2**, 110–111 (2000)
33. Kihara, T., Liu, X.Y., et al.: Direct electron transfer to hydrogenase for catalytic hydrogen production using a single-walled carbon nanotube forest. Int. J. Hydrogen Energy **36**, 7523–7529 (2011)
34. Klerke, A., Christensen, C.H., et al.: Ammonia for hydrogen storage: challenges and opportunities. J. Mater. Chem. **18**, 2304–2310 (2018)
35. Kohlmann, C., Greiner, L., et al.: Ionic liquids as performance additives for electroenzymatic syntheses. Chem. Eur. J. **15**, 11692–11700 (2009)
36. Kohlmann, C., Lütz, S.: Electroenzymatic synthesis of chiral sulfoxides. Eng. Life Sci. **6**, 170–174 (2006)
37. Kohlmann, C., Lütz, S.: Electroenzymatic synthesis. In: Hammerich, O. (ed.) Organic Electrochemistry, 5th edn., pp. 1511–1542. CRC Press, New York (2015)
38. Krassen, H., Stripp, S., et al.: Immobilization of the [FeFe]-hydrogenase CrHydA1 on a gold electrode: design of a catalytic surface for the production of molecular hydrogen. J. Biotechnol. **142**, 3–9 (2009)
39. Krieg, T., Huttmann, S., et al.: Gas diffusion electrode as novel reaction system for an electro-enzymatic process with chloroperoxidase. Green Chem. **13**, 2686–2689 (2011)
40. Kuk, S.K., Gopinath, K., et al.: NADH-free electroenzymatic reduction of CO_2 by conductive hydrogel-conjugated formate dehydrogenase. ACS Catal. **9**, 5584–5589 (2019)
41. Kuwabata, S., Tsuda, R., et al.: Electrochemical conversion of carbon dioxide to methanol with the assistance of formate dehydrogenase and methanol dehydrogenase as biocatalysts. J. Am. Chem. Soc. **116**, 5437–5443 (1994)
42. Laane, C., Weyland, A., Franssen, M.: Bioelectrosynthesis of halogenated compounds using chloroperoxidase. Enzyme Microb. Technol. **8**, 345–348 (1986)
43. Lee, K., Moon, S.H.: Electroenzymatic oxidation of veratryl alcohol by lignin peroxidase. J. Biotechnol. **102**, 261–268 (2003)
44. Ley, C., Schewe, H., et al.: Coupling of electrochemical and optical measurements in a microtiter plate for the fast development of electro enzymatic processes with P450s. J. Mol. Catal. B Enzym. **92**, 71–78 (2013)
45. Lukoyanov, D., Khadka, N., et al.: Reductive elimination of H_2 activates nitrogenase to reduce the N≡N triple bond: characterization of the $E_4(4H)$ Janus intermediate in wild-type enzyme. J. Am. Chem. Soc. **138**, 10674–10683 (2016)
46. Lütz, S., Steckhan, E., Liese, A.: First asymmetric electroenzymatic oxidation catalyzed by a peroxidase. Electrochem. Commun. **6**, 583–587 (2004)
47. Lütz, S., Vuorilehto, K., Liese, A.: Process development for the electroenzymatic synthesis of (R)-methylphenylsulfoxide by use of a 3-dimensional electrode. Biotechnol. Bioeng. **98**, 525–534 (2007)
48. Mazurenko, I., Monsalve, K., et al.: Impact of substrate diffusion and enzyme distribution in 3D-porous electrodes: a combined electrochemical and modelling study of a thermostable H_2/O_2 enzymatic fuel cell. Energy Environ. Sci. **10**, 1966–1982 (2017)
49. McDonald, T.J., Svedruzic, D., et al.: Wiring-up hydrogenase with single-walled carbon nanotubes. Nano Lett. **7**, 3528–3534 (2007)
50. Milton, R.D., Abdellaoui, S., et al.: Nitrogenase bioelectrocatalysis: heterogeneous ammonia and hydrogen production by MoFe protein. Energy Environ. Sci. **9**, 2550–2554 (2016)
51. Milton, R.D., Cai, R., et al.: Bioelectrochemical haber-bosch process: an ammonia-producing H_2/N_2 fuel cell. Angew. Chem. Int. Ed. **56**, 2680–2683 (2017)
52. Milton, R.D., Minteer, S.D.: Nitrogenase bioelectrochemistry for synthesis applications. Acc. Chem. Res. **52**, 3351–3360 (2019)
53. Monsalve, K., Mazurenko, I., et al.: Impact of carbon nanotube surface chemistry on hydrogen oxidation by membrane-bound oxygen-tolerant hydrogenases. Chem. ElectroChem. **3**, 2179–2188 (2016)

54. Morra, S., Valetti, F., et al.: Hydrogen production at high faradaic efficiency by a bio-electrode based on TiO_2 adsorption of a new [FeFe]-hydrogenase from *Clostridium perfringens*. Bioelectrochemistry **106**, 258–262 (2015)
55. Oh, Y.K., Jin Lee, Y., et al.: Bioelectrocatalytic hydrogen production using Thiocapsa roseopersicina hydrogenase in two-compartment fuel cells. Int. J. Hydrogen Energy **33**, 5218–5223 (2008)
56. Palmore, G.T.R., Bertschy, H.: A methanol/dioxygen biofuel cell that uses NAD^+-dependent dehydrogenases as catalysts: application of an electro-enzymatic method to regenerate nicotinamide adenine dinucleotide at low overpotentials. Electroanal. Chem. **443**, 155–161 (1998)
57. Pankratov, D., Conzuelo, F., et al.: A nernstian biosupercapacitor. Angew. Chem. **55**, 15434–15438 (2016)
58. Parkinson, B.A., Weaver, P.F.: Photoelectrochemical pumping of enzymatic CO_2 reduction. Nature **309**, 148–149 (1984)
59. Pletcher, D.: Indirect oxidations using electrogenerated hydrogen peroxide. Acta Chem. Scand. **53**, 745–750 (1999)
60. Reda, T., Plugge, M.C., et al.: Reversible interconversion of carbon dioxide and formate by an electroactive enzyme. Proc. Natl. Acad. Sci. USA **105**, 10654–10658 (2008)
61. Risbridger, T., Anderson, R., et al.: Plastic waste as a feedstock for solar-driven H_2 generation. In: RSC Energy Environmental Series Royal Society of Chemistry, pp. 17–62 (2018)
62. Robinson, W.E., Bassegoda, A., et al.: Oxidation-state-dependent binding properties of the active site in a Mo-containing formate dehydrogenase. J. Am. Chem. Soc. **139**, 9927–9936 (2017)
63. Rohde, M., Sippel, D., et al.: The critical E_4 state of nitrogenase catalysis. Biochemistry **57**, 5497–5504 (2018)
64. Ruinatscha, R., Buehler, K., Schmid, A.: Development of a high performance electro-chemical cofactor regeneration module and its application to the continuous reduction of FAD. J. Mol. Catal. B Enzym. **103**, 100–105 (2014)
65. Sagara, T., Hirayama, K.: Electrode-reaction of the soluble domain of the membrane-bound hydrogenase from desulfovibrio-vulgaris, strain miyazaki-f. Bioelectrochem. Bioenerg. **28**, 191–204 (1992)
66. Sakai, K., Kitazumi, Y., et al.: Direct electron transfer-type four-way bioelectrocatalysis of CO_2/formate and NAD^+/NADH redox couples by tungsten-containing formate dehydrogenase adsorbed on gold nanoparticle-embedded mesoporous carbon electrodes modified with 4-mercaptopyridine. Electrochem. Commun. **84**, 75–79 (2017)
67. Schlager, S., Fuchsbauer, A., et al.: Carbon dioxide conversion to synthetic fuels using biocatalytic electrodes. J. Mater. Chem. A **5**, 2429–2443 (2017)
68. Schmitz, L.M., Rosenthal, K., Lütz, S.: Enzyme-Based Electrobiotechnological Synthesis. In: Harnisch, F., Holtmann, D., (eds) Bioelectrosynthesis. Advances in Biochemical Engineering/Biotechnology, p. 167. Springer (2017)
69. Seefeldt, L.C., Hoffman, B.M., et al.: Energy transduction in nitrogenase. Acc. Chem. Res. **51**, 2179–2186 (2018)
70. Seelbach, K., van Deurzen, M.P.J., et al.: Improvement of the total turnover number and space-time yield for chloroperoxidase catalyzed oxidation. Biotechnol. Bioeng. **55**, 283–288 (1997)
71. Shiraiwa, S., So, K., et al.: Reactivation of standard [NiFe]-hydrogenase and bioelectrochemical catalysis of proton reduction and hydrogen oxidation in a mediated-electron-transfer system. Bioelectrochemistry **123**, 156–161 (2018)
72. Sode, K., Yamazaki, T., et al.: BioCapacitor: a novel principle for biosensors. Biosens. Bioelectron **76**, 20–28 (2016)
73. Sokol, K.P., Robinson, W.E., et al.: Photoreduction of CO_2 with a formate dehydrogenase driven by photosystem II using a semi-artificial Z-scheme architecture. J. Am. Chem. Soc. **140**, 16418–16422 (2018)
74. Sokol, K.P., Robinson, W.E., et al.: Reversible and selective interconversion of hydrogen and carbon dioxide into formate by a semiartificial formate hydrogenlyase Mimic. J. Am. Chem. Soc. **141**, 17498–17502 (2019)

75. Sono, M., Roach, M.P., et al.: Heme-containing oxygenases. Chem. Rev. **96**, 2841–2888 (1996)
76. Steckhan, E.: Electroenzymatic synthesis. Top. Curr. Chem. **170**, 83–111 (1994)
77. Steckhan, E., Arns, T., et al.: Environmental protection and economization of resources by electroorganic and electroenzymatic syntheses. Chemosphere **43**, 63–73 (2001)
78. Strohle, F.W., Cekic, S.Z., et al.: A computational protocol to predict suitable redox mediators for substitution of NAD(P)H in P450 monooxygenases. J. Mol. Catal. B Enzym. **88**, 47–51 (2013)
79. Svedružić, D., Blackburn, J.L., et al.: Hydroalkoxylation catalyzed by a gold(I) complex encapsulated in a supramolecular host. J. Am. Chem. Soc. **133**, 4299–4306 (2011)
80. Tosstorff, A., Dennig, A., et al.: Mediated electron transfer with monooxygenases-insight in interactions between reduced mediators and the cosubstrate oxygen. J. Mol. Catal. B Enzym. **108**, 51–58 (2014)
81. Udit, A.K., Arnold, F.H., Gray, H.B.: Cobaltocene-mediated catalytic monooxygenation using holo and heme domain cytochrome P450BM3. J. Inorg. Biochem. **98**, 1547–1550 (2004)
82. van de Velde, F., van Rantwijk, F., Sheldon, R.A.: Selective oxidations with molecular oxygen, catalyzed by chloroperoxidase in the presence of a reductant. J. Mol. Catal. B Enzym. **6**, 453–461 (1999)
83. Varfolomeyev, S.D., Yaropolov, A.I., et al.: Bioelectrocatalysis—The electrochemical kinetics of hydrogenase action. J. Biotechnol. **27**, 331–339 (1993)
84. Xiao, X., Xia, H.Q., et al.: Tackling the challenges of enzymatic (Bio)fuel cells. Chem. Rev. **119**, 9509–9558 (2019)
85. Yang, Z.Y., Ledbetter, R., et al.: Evidence that the P_i release event is the rate-limiting step in the nitrogenase catalytic cycle. Biochemistry **55**, 3625–3635 (2016)
86. Yaropolov, A.I., Karyakin, A.A.: Oxidation-reduction properties of several low potential iron-sulfur proteins and of methylviologe. Bioelectrochemistry Bioenerg. **12**, 267–277 (1984)
87. Yu, E.H., Scott, K.: Enzymatic biofuel cells-fabrication of enzyme electrodes. Energies **3**, 23–42 (2010)
88. Yuan, M., Sahin, S., et al.: Creating a low-potential redox polymer for efficient electroenzymatic CO_2 reduction. Angew. Chem. Int. Ed. **57**, 6582–6586 (2018)
89. Trost, E-M., Fischer, L.: Minimization of by-product formation during amino acid oxidase catalyzed racemate resolution of amino acids. J Mol Catal B Enzym **19–20**:189–195 (2002).
90. Buto, S., Pollegioni, L., D'Angiuro, L., Pilone, M.S.: Evaluation of D -amino-acid oxidase from Rhodotorula gracilis for the production of α- keto acids: a reactor system. Biotechnol Bioeng **44**:1288–1294 (1994).
91. Kawabata, S., Iwata, N., Yoneyama, H.: Asymmetric electrosynthesis of amino acid using an electrode modified with amino acid oxidase and electron mediator. Chem Lett **2**:110–111 (2000).

Bioelectrosynthesis of Value-Added Compound Production

Jessica Crivelaro Pacheco, Graziela Cristina Sedenho, and Frank N. Crespilho

Abstract The bioelectrosynthesis of value-added compounds has attracted significant attention in the field of bioelectrochemistry. Current industrial processes generally have high energy consumption and require a high supply of non-renewable catalysts and feedstocks, resulting in a poor efficiency and sustainability. In this context, bioelectrochemical systems are considered a promising alternative to some traditional industrial processes, such as the Haber–Bosch process, which produces NH_3 from N_2 and H_2. The Haber–Bosch process accounts for 2% of the global energy consumed, while it is also responsible for 3% of CO_2 emissions worldwide. Bioelectrochemical systems are also employed in processes of economic and environmental interest such as the capture of CO_2 and its conversion into value-added compounds and the production of H_2 fuel. This chapter will therefore address the latest advances in the bioelectroreduction of N_2 to NH_3, CO_2 to formate, and the bioelectrosynthesis of H_2. The enzymatic mechanisms and main methods of enzyme immobilization and stabilization to achieve high-efficiency bioelectrochemical systems are also described.

1 Introduction

The use of redox enzymes for the bioproduction of value-added compounds has attracted significant attention in modern bioelectrochemistry. This includes the transformation of atmospheric CO_2 into fuels and chemicals such as methanol, formaldehyde, and formate [1, 2]. Another abundant atmospheric gas that can be employed in

J. C. Pacheco · G. C. Sedenho · F. N. Crespilho (✉)
São Carlos Institute of Chemistry, University of São Paulo (USP), São Carlos, SP 13560-970, Brazil
e-mail: frankcrespilho@iqsc.usp.br

J. C. Pacheco
e-mail: jessica.pacheco@usp.br

G. C. Sedenho
e-mail: graziela.cs@usp.br

F. N. Crespilho (ed.), *Advances in Bioelectrochemistry Volume 4*,
https://doi.org/10.1007/978-3-030-99662-8_3

the bioproduction of value-added compounds is N_2, which can be converted to NH_3 or chiral amino acids, which in turn are used as precursors in organic syntheses and pharmaceutical industries [3, 4]. A final example of an important bioelectrocatalytic process is the electrosynthesis of H_2. From an environmental perspective, H_2 is a promising alternative to fossil fuels because it is a clean fuel and is associated with high energy content.

The electroreduction of CO_2 is a process with potentially great economic and environmental impact because it allows for the mitigation of the atmospheric emissions of this pollutant, while simultaneously producing fuels and other useful products [5, 6]. However, CO_2 is a thermodynamically stable molecule, and a significant energy barrier must be overcome to reduce this molecule. For this reaction to occur, high pressures, high overpotentials, or the use of catalysts are required. Significant efforts have been made to develop catalysts to overcome the thermodynamic and kinetic limitations associated with the chemical activation of CO_2. However, producing a catalyst with a favorable cost–benefit ratio, good selectivity, and efficiency to reduce CO_2 to a single product, and which requires low overpotentials, remains a significant challenge [7].

The reduction of N_2 to NH_3, while also being a reaction of great economic and environmental interest, is another process associated with high demands of energy. NH_3 is widely used in agriculture and is also employed as a substrate in the synthesis of chiral amino acids, which are important precursors in food, agricultural, pharmaceutical, and organic synthesis industries [8]. The Haber–Bosch process produces 500 million tons of NH_3 annually, from N_2 and H_2 gases. This is estimated to account for 2% of the global energy consumed and for 3% of global CO_2 emissions [9]. The importance of identifying economically viable and sustainable alternative routes to reduce N_2 is therefore evident.

The electrosynthesis of H_2 is another important electrocatalytic process. H_2 has an energy content that is 2.7 times that of traditional hydrocarbon fuels such as gasoline [10]. Ironically, most routes for H_2 synthesis currently use these hydrocarbons as substrates [11]. Another alternative to obtain H_2 gas is to cleave water molecules. This process is however relatively energy-intensive and requires suitable catalysts. Pt has traditionally been used as electrocatalyst for this purpose, and although it is highly efficient and enables the production of high levels of H_2 at negligible electrochemical overpotential; it is a rare and highly expensive metal, which limits its use in a large scale [12].

Despite several advances made in the processes discussed above, the development of sustainable alternative strategies remains of significant interest to scientists globally. In this context, redox enzymes represent promising alternative biocatalysts for these reactions, while contributing to the evolution in the field of synthesis of value-added compounds [13, 14]. In addition, redox enzymes offer several advantages over conventional inorganic catalysts because they can operate under mild conditions of temperature, pressure, and pH; show high specificity and efficiency; and are renewable and widely available. Moreover, redox enzymes can provide alternative strategies for the synthesis of more efficient biocatalysts [15].

Among the redox enzymes, nitrogenases, formate dehydrogenases, and hydrogenases have been regarded as the most efficient biocatalysts for the reactions in discussion. Nitrogenases are capable of producing NH_3 from N_2 under mild pH, temperature, and pressure conditions [4]. Formate dehydrogenases can reduce CO_2 to formate, which is the first stable intermediary in the reduction of CO_2 to methanol or methane, and can also be used as a fuel or as a raw material in the chemical industry [16]. Finally, hydrogenases promote the efficient reduction of protons to H_2 [17]. In this context, the present chapter provides an overview of the bioelectrosynthesis of value-added compounds from atmospheric CO_2 and N_2 and the production of H_2 fuel.

2 Bioelectroreduction of N_2 to NH_3 by Nitrogenases

Biological fixation of N_2 occurs naturally in diazotrophic microorganisms through nitrogenases [18]. Like most enzymes, nitrogenase operates at room temperature and atmospheric pressure, which presents a significant advantage when compared to traditional catalysts used in the Haber–Bosch process. The reduction of N_2 to NH_3 by nitrogenases follows equation (1) under ideal conditions, which is illustrated as follows:

$$N_2 + 8H^+ + 16MgATP + 8e^- \rightarrow 2NH_3 + H_2 + 16MgADP + 16Pi \quad (1)$$

Among the nitrogenases, Mo-dependent nitrogenase from *Azotobacter vinelandii* (PDB ID: 4WZA), *Klebsiella pneumoniae* (PDB ID: 1QH8), and *Clostridium pasteurianum* (PDB ID: 4WES) are the most abundant and widely studied enzymes [19]. MoFe nitrogenase has two protein components, i.e., the MoFe protein and the Fe protein (FeP) that work together to reduce N_2 to NH_3 [9] (Fig. 1). The MoFe protein is an $\alpha_2\beta_2$-tetramer with a molecular weight of ~220 kDa, and its α- and β-subunits are encoded by *nifD* and *nifK*, respectively. The MoFe protein also contains the active site [7Fe-9S-1Mo-C-homocitrate] (FeMo–co), where substrate reduction occurs, and a [8Fe-7S] P cluster that functions as an electron carrier. The Fe protein, encoded by *nifH*, is a homodimer with a molecular weight of ~60 kDa and only has one cluster [4Fe–4S] that transfers an electron to the MoFe protein during the transient association of the MoFe and Fe proteins [20]. In the catalytic cycle, the Fe protein is initially reduced by either ferredoxin or flavodoxin, resulting in the hydrolysis of two MgATP equivalents for each electron generated [3]. One electron is then transferred from the FeP to the P cluster and finally to the FeMo-co (Fig. 1).

In FeMo-co, the electrons are trapped in the form of hydrides around the Fe centers, before being used in N_2 reduction through a complex process incorporating eight individual states (E_0–E_7), as illustrated in the Thorneley-Lowe scheme (Fig. 2) [21]. The catalytic mechanism of N_2 reduction does present some points to consider, such as the ability of FeP to transfer four consecutive electrons, all at the same redox potential, to FeMo-co while the metal cofactor is in its ground state, E_0. Another

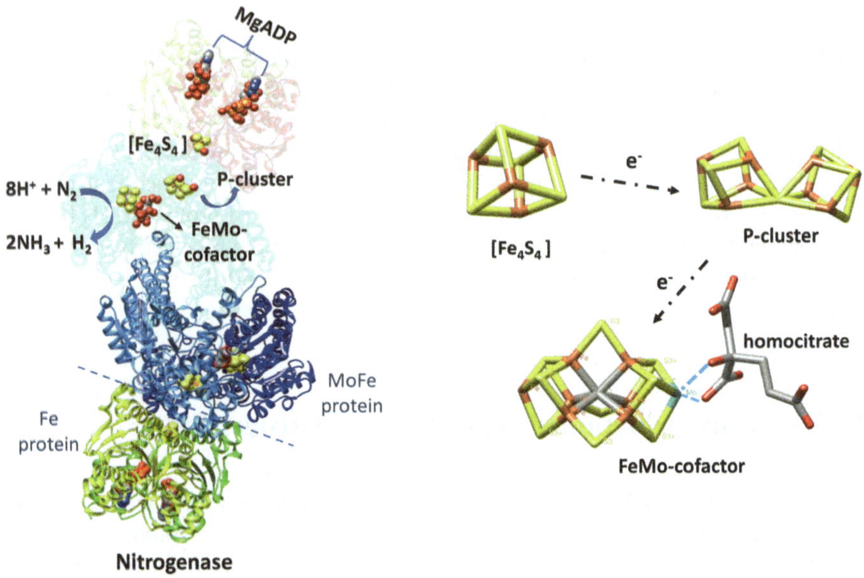

Fig. 1 Nitrogenase MoFe protein from *Azotobacter vinelandii* (PDB ID: 1M1N) and the essential metalloclusters: [Fe_2S_4], P cluster, and FeMo-cofactor, where N_2 is reduced. Atoms are colored as follows: S, yellow; Fe, orange; C, gray; Mo, cyan; N, blue; and O, red. The image was generated with Chimera

unknown aspect is the time between the two electron-transfer events, which may be expected to be time-consuming and would require that all the intermediates are sufficiently stable [21]. The most widely accepted hypothesized mechanism is that each transferred electron pair is stored as a hydride (H^-) in the metal surface [21], in which case only one electron would reside in the cofactor itself, as represented by the odd-numbered E states of the early stages of the Thorneley-Lowe scheme (Fig. 2). In the E_2 state, the first surface hydride is formed, followed by a second in the E_4 state. At this point, these two hydrides recombine into one H_2 molecule. The H_2 then rapidly moves away from the cofactor, leaving the active site in a high-energy, super-reduced state [21], where the nitrogenase subsequently activates N_2, breaking the stable triple bond, and initiating biological N_2 fixation to form two NH_3 molecules following three more electron-transfer states (E_5, E_6, and E_7, Fig. 2) [22].

Despite the proposed mechanism (Fig. 2), it is still not clear as to how the protons/electrons, or hydride species are involved in the cleavage of the N_2 bond (second half of the Thorneley and Lowe scheme). Two potential pathways proposed for the hydrogenation of N_2 are the distal and alternating pathways (Fig. 3), where N_2 would bind to the metallic center (M–N_2) of FeMo-co and eventually form a species with a terminal amine (M–NH_2). In the final step, the transfer of a proton/electron equivalent to a terminal starch intermediate releases NH_3 and regenerates the E_0. The difference between the two mechanisms proposed is the protonation site of N_2. Computational studies have pointed out that potential pathways differ at the metal

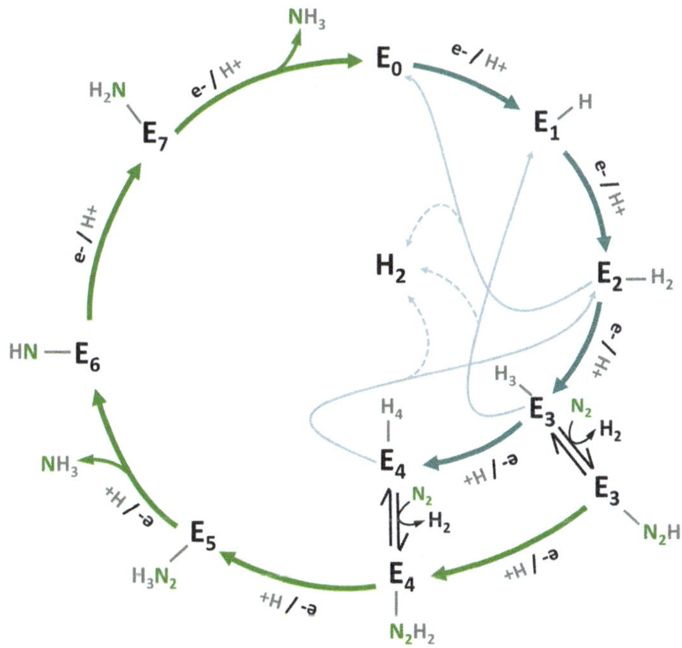

Fig. 2 Thorneley and Lowe scheme, in which the eight steps of reduction (E_0–E_7) represent the enzymatic catalysis of nitrogenase

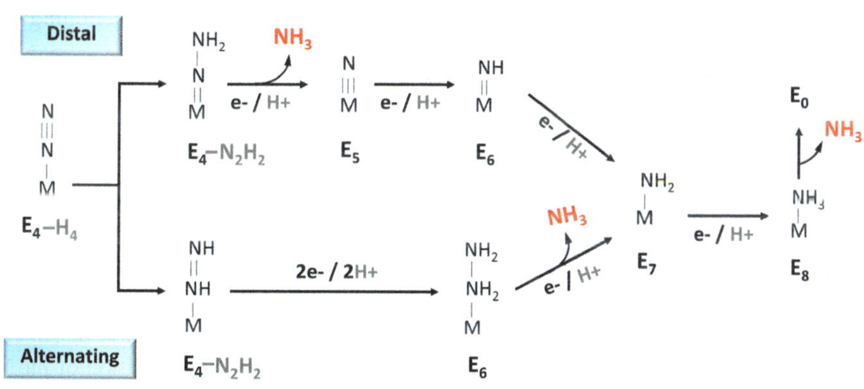

Fig. 3 Distal (upper) and alternate (lower) N_2 reduction mechanism. The state E_n (E_4–E_8) corresponding to the Thorneley-Lowe model is indicated at each intermediate along with the proposed point of NH_3 release

ion associated with the active center, in other words, the point where the reaction takes place [23]. For the distal route, Mo is involved in the reaction where a single N atom of N_2 is hydrogenated in three steps, at which point the first NH_3 is released. The remaining N atom (E_5, Fig. 3) is then hydrogenated thrice before the second NH_3 is released [24]. In the alternating pathway, the Fe from FeMo-co participates in a reaction where the two N atoms are alternately hydrogenated in four hydrogenation steps, generating a hydrazine-bound state, and releasing the first NH_3 only in the fifth step [25]. Although studies of intermediates of N_2 reduction by nitrogenase indicate that the alternating pathway is the most likely [26], no scientific consensus has been reached on the correct pathway [27].

Although the catalytic kinetics of the MoFe protein has been widely studied, the thermodynamic driving forces responsible for its exceptional reactivity remain poorly understood [28]. The complexity of the nitrogenases process makes studies to elucidate the mechanism of electron transfer and ATP coupling challenging. In addition, the potentially limited electron transfer between the two protein components of the nitrogenase enzyme hampers the understanding of the electron-transfer mechanisms inside MoFe nitrogenases [29]. The cascade of bi-enzymes further requires a dynamic interaction between the two proteins, i.e., they must come together to transfer electrons, separate to consume ATP, and then unite again to transfer electrons again. ATP regeneration is further required, as the enzyme is ATP-dependent. This ATP regeneration system complicates the functioning of the electrochemical cell because the system consists of more than one enzyme. The Fe protein further requires 16 ATP molecules to transfer the eight electrons required for the high-energy enzymatic synthesis of NH_3. This also presents a challenge for the functioning of the electrochemical cell [28].

Bioelectrocatalytic studies have so far mostly focused on the electrochemical control of the MoFe protein, where the Fe protein and ATP are not involved. Recently, the ATP-independent, direct bioelectrocatalytic reduction of N_2 to NH_3 by MoFe nitrogenases has been studied. MoFe nitrogenases were successfully immobilized and stabilized into a hydrogel matrix on the surface of a carbon electrode [30]. The electrochemical results further indicated that the P cluster and the FeMo-co could interact independently with the electrode. This represents intriguing, new opportunities for mechanistic studies aimed at determining the mechanism of electron transfer with the substrate and the role of each cofactor [30]. The incorporation of MoFe nitrogenase onto the hydrogel matrix further allowed for the determination of the P cluster and FeMo-co redox potentials, which were -0.23 ± 0.01 and -0.60 ± 0.01 V, respectively [29]. It was also possible to verify that the bioelectrochemical catalysis only occurred in the presence of <1.8% H_2, and in the absence of FeP. Here, the redox potential of FeMo-co was shifted to -0.43 V in the absence of H_2. A similar redox potential change was observed when FeP was incorporated into the enzyme electrode in the presence of H_2. These observations indicated that the change in the redox potential of FeMo-co occurred through the interaction with H_2 and that FeP potentially prevents this interaction. Alternatively, an endergonic electron-transfer sequence during the catalytic turnover of nitrogenase has also been proposed. As the P cluster is less reductive compared to FeMo-co, the electron transfer from the P cluster to FeMo-co would require energy. Within the nitrogenase catalytic cycle, this process would then occur through a series of consecutive, combined proton–electron transfers. These transfers are driven by the formation of Fe-hydride bonds in FeMo-co. Based on a study of the redox potential of the P cluster as a function of pH, it was proposed that the electrochemical catalysis in this system occurs in a similar manner as the native nitrogenases mechanism. It may therefore be reasonable to assume that the electrochemically driven MoFe turnover occurs via sequential, proton-coupled electron-transfer steps.

The increase in the electron-transfer rate using artificial redox mediators is a promising strategy to increase NH_3 production in industrial bioelectrochemical processes. Electrochemical studies have shown that the use of cobaltocene [31] and viologen [32] derivatives as redox mediators resulted in higher Faradaic efficiency for NH_3 bioelectrosynthesis than in the absence of the redox mediators. The use of redox mediators, such as cobaltocene derivatives, further provided efficient electrical wiring of nitrogenase even in the absence of a reducing protein (FeP) and ATP (which is a rate-limiting step for the overall catalysis). The electron transfer was independent of the distance from the electrode surface and the enzyme orientation, providing increased efficiency toward NH_3 production (Fig. 4). Therefore, this is a promising strategy for industrial and large-scale NH_3 bioelectrosynthesis [31].

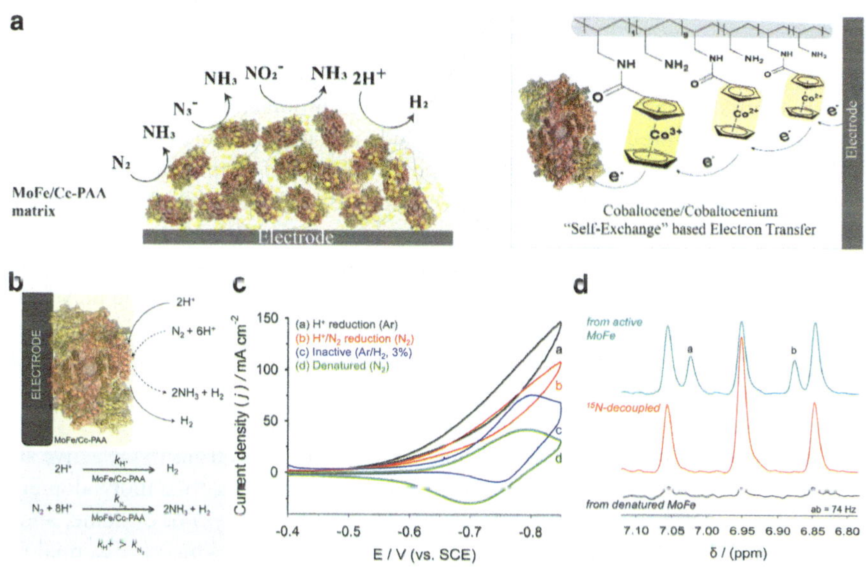

Fig. 4 Schematic of **a** the MoFe/Cc-PAA electrode and of **b** the bioelectrochemical potential reaction of MoFe protein turnover. **c** cyclic voltammograms of MoFe/Cc-PAA under different conditions. **d** ^{15}N-non-decoupled and ^{15}N-decoupled ^{1}H NMR of $^{15}NH_3$ produced from bulk electrolysis of $^{15}N_2$ using MoFe/Cc-PAA. Reprinted with permission from [31]. Copyright 2021 American Chemical Society

The reaction of N_2 reduction by nitrogenases can also be the first step in an enzymatic cascade that produces chiral amino acids at high reaction rates. Such chiral amino acids are of great interest because they are often used as precursors in agricultural and pharmaceutical industries and as chiral targeting aids and chiral synthons in organic syntheses [8]. Bioelectrosynthesis of NH_3 from N_2 by nitrogenases is therefore of great importance for the production of other value-added compounds.

3 Bioelectroreduction of CO_2 by Redox Enzymes

Several value-added compounds, such as methane, methanol, and formate, and other fuels can be produced from the catalytic reduction of CO_2 [33, 34]. In particular, formate has attracted attention because it is the first stable intermediary in the reduction of CO_2 to methanol or methane and can also be used as a raw material in the chemical industry as well as an intermediate in energy-conserving pathways. Further, formate is non-flammable, non-toxic, and moderately inert. This offers significant advantages in terms of its storage and transportation, specifically when compared to other fuels with similar redox potentials, such as H_2 [35].

Due to the high stability of CO_2 molecules, significant efforts have been made to develop catalysts to overcome the thermodynamic and kinetic challenges associated with the chemical activation of CO_2 [16, 36–38]. In addition, potential catalysts show a satisfactory cost–benefit ratio. In this context, the use of redox enzymes as biocatalysts has proven to be an interesting and promising strategy. Among the enzymes capable of reducing CO_2 to value-added products, formate dehydrogenase (FDH) is most widely studied. This enzyme is abundant in nature and catalyzes the interconversion of CO_2 to $HCOO^-$ via two electrons [39]. FDHs can be classified as either metal-dependent (MD) or metal-independent (MI) [40]. Most MI-FDHs are stable in air and are active over a wide pH range but are not inhibited or deactivated by O_2. However, their catalytic activity is lower than that of MD-FDHs, and they favor the oxidation of $HCOO^-$ rather than the reduction of CO_2 [1]. In addition, their dependence on nicotinamide adenine dinucleotide phosphate oxidase (NAD(P)H) limits enzyme immobilization at the electrode surface [41]. MD-FDHs are therefore considered more suitable for reducing CO_2 than MI-FDHs are.

MD-FDHs contain FeS clusters capable of transferring electrons to the active site [42]. This active site contains a Mo or W atom coordinated to two molybdopterin guanine dinucleotides (MGD), a selenocysteine residue (SeCys) or cysteine, and a terminal sulfide ligand (=S/−SH) introduced by a chaperone, which is essential for the activity of the enzyme (Fig. 5) [42].

The catalytic reduction of carbon dioxide by Mo-dependent FDHs occurs when carbon dioxide binds to the reduced active site of the enzyme, while a protonated sulfo group, Mo^{4+}–SH, remains protonated. CO_2 is then reduced through the direct transfer of hydride from the protonated sulfo group of the reduced Mo center, Mo^{4+}–SH, producing formate and $Mo^{6+}=S$. The reduction of Mo^{6+}–Mo^{4+} completes the catalytic cycle via the intramolecular transfer of electrons from another redox

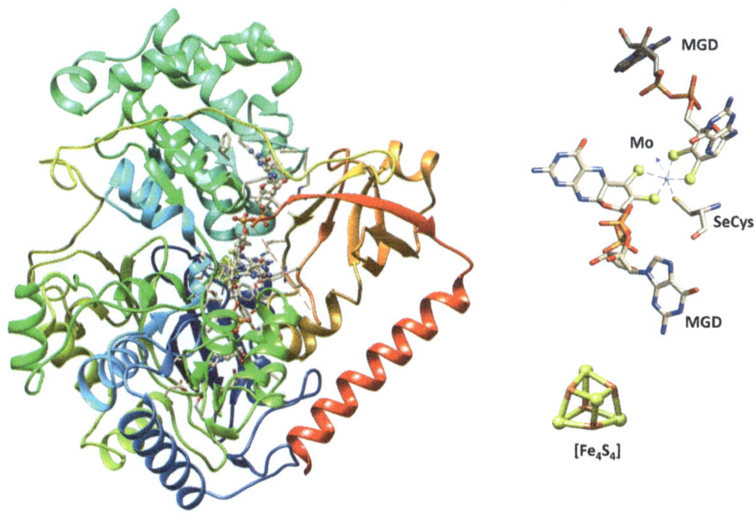

Fig. 5 Structure of an MD-FDH formate dehydrogenase obtained from *Escherichia coli* (*Ec*FDH-H, PDB ID: 1AA6). The essential metalloclusters are shown, i.e., the $[Fe_4S_4]$ cluster and the active site that contains a Mo atom coordinated by two molybdopterin guanine dinucleotides (MGD) and a selenocysteine residue (SeCys). Atoms are colored as follows: S, yellow; Fe, orange; C, gray; Mo, cyan; N, blue; and O, red. The image was generated with Chimera

center(s) before the formate is released (Fig. 6). Since the reduced Mo^{4+}favors the protonation of the sulfo group, the (Mo^{4+}–SH) active site is subsequently regenerated (Fig. 6). In this mechanism, the CO_2 binds to the same location as the formate (in the reverse mechanism). Furthermore, arginine is conserved, anchoring its oxygen atom(s) through hydrogen bonds, while orienting its carbon atom toward the protonated sulfo ligand [33].

Despite a significant number of studies, the FDH mechanism to oxidize formate, or reduce CO_2, is still not fully understood [43]. One thing that is not clear at this stage is if the SeCys group dissociates from the metal during catalysis, being replaced by formate or CO_2, or the metal coordination is maintained. This should be further investigated [44].

Since inhibitor binding to the active site can provide important information about metal coordination, the inhibition mechanism of FDH-H from *Escherichia coli* (*Ec*FDH-H, PDB ID: 1AA6) activity was subsequently investigated using protein film electrochemistry. Previous studies have shown that formate oxidation is strongly and competitively inhibited by the presence of N_3^-, NO_2^-, NO_3^-, OCN^-, and SCN, while CO_2 reduction is weakly and not competitively inhibited (Fig. 7). Since interconversion between $Mo^{6+} = S$ and Mo^{4+}–SH occurs at the active site during catalysis, the catalytic reaction may be expected to be inhibited by the binding of an inhibitor to the $Mo^{6+} = S$ state. Overall, the clear relationship between inhibitor binding affinity to the $Mo^{6+} = S$ state and the electron-donating strength of the inhibitor suggests

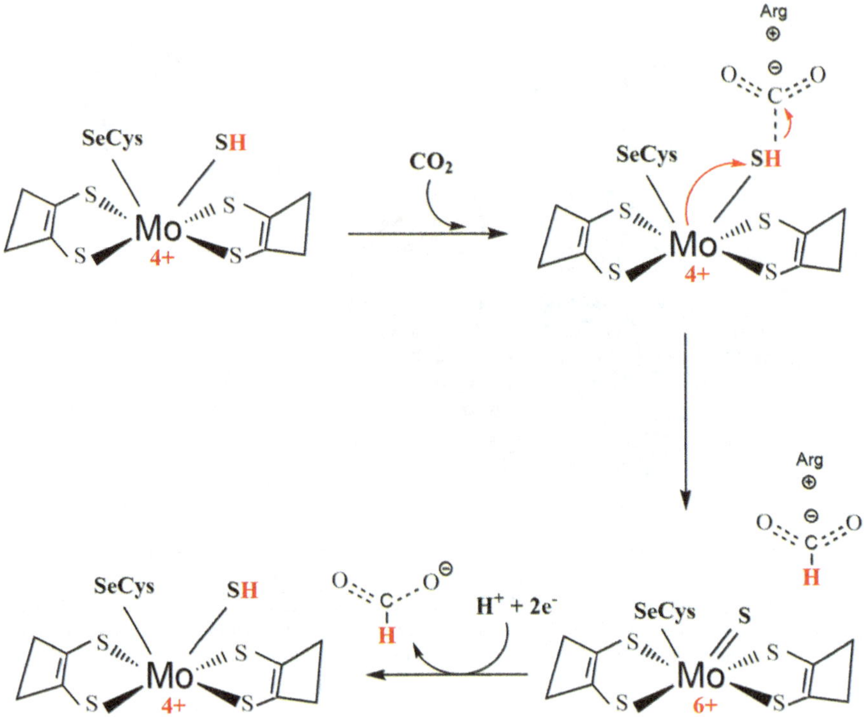

Fig. 6 Mechanism of CO_2 reduction by Mo-dependent FDH

that the inhibitors or substrates bind directly to the metal center. This study therefore suggests dissociation of the selenocysteine ligand during catalysis [44]. These results, indicating that these inhibitor agents can indeed inhibit the reverse reaction, are important for the production of value-added compounds from CO_2, since this could ultimately lead to a higher efficiency of formate bioelectrosynthesis.

In addition to FDH from *Escherichia coli,* other Mo-dependent FDHs were also shown to be able to interconvert CO_2/formate when they were immobilized on electrode surfaces. These FDHs included those obtained from *Desulfovibrio desulfuricans* (*Dd*FDH) [38], W-dependent FDHs from *Syntrophobacter fumaroxidans* (*Sf*FDH) [42], *Desulfovibrio vulgaris Hildenborough* (*Dv*H-FDH, PDB ID: 6SDV) [45], and *Clostridium ljungdahlii* (*Cl*FDH) [46]. The enzymes *Cl*FDH [46], *Dd*FDH [47], and *Ec*FDH-H [48] did show a preference for formate oxidation, with an associated low reduction of CO_2. Despite the low CO_2 reduction, important information regarding the enzymatic mechanism of FDHs was identified [47]. This included confirmation that formate oxidation and CO_2 reduction occur through the transfer of a sulfo group hydride from either the oxidized or reduced Mo center, $Mo^{6+}=S$ and Mo^{4+}–SH, respectively. The reduction potentials associated with the Mo (VI/IV), Mo (VI/V), and Mo (V/IV) pairs and the number of electrons involved in the catalysis reaction were also determined [38].

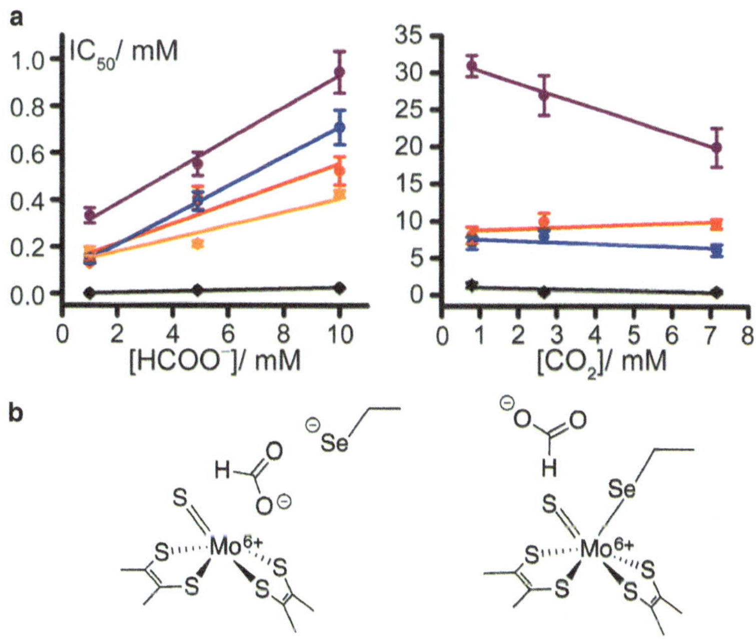

Fig. 7 **a** Inhibitor IC_{50} values as a function of substrate concentration. NO_2^- was reduced by the electrode and was omitted from the graph. $N3^-$, black; OCN^-, red; SCN^-, blue; NO_3^-, purple; and NO_2^-, orange. **b** Formate approaching a vacant coordination site on Mo(VI) and a coordination sphere where Se bonded with the Mo(VI) center. Reprinted with permission from [44]. Copyright 2021 American Chemical Society

FDH from *Rhodobacter aestuarii* (*Ra*FDH), a newly discovered NAD-linked enzyme that contains Mo, was also evaluated for CO_2 reduction and formate oxidation along with its turnover rate [49]. Here, CO_2 reduction by *Ra*FDH was much more efficient than formate oxidation. This is unprecedented for dehydrogenases formate, as most enzymes show high rates of formate oxidation. The high CO_2 reducing efficiency of *Ra*FDH combined with the less prominent inverse reaction of formate oxidation present a promising strategy to obtain an efficient electro-enzymatic system for the production of formate from CO_2. *Ra*FDH does however show lower turnover rates for CO_2 reduction than those of other enzymes. *Sf*FDH can, for instance, reduce CO_2 with a higher turnover rate than that of *Ra*FDH, but its formate oxidation rate is 12 times than of CO_2 reduction [42]. Currently, the major challenge for the practical application of FDHs for CO_2 reduction would therefore be to combine a high turnover rate for CO_2 reduction with low rates of formate oxidation.

4 H_2 Gas Production by Hydrogenases

Another important bioelectrocatalytic process is the electrosynthesis of H_2 fuel from protons employing hydrogenases. Hydrogenase was first described in 1932 by Stephenson and Stickland [50]. This unique enzyme is known for catalyzing the reversible production of H_2 at low overpotential, with efficiency rates comparable to Pt [17, 50]. Studies aimed at understanding its mechanism, and the optimization of its use in the production of H_2, employed strategies ranging from immobilization to genetic engineering [51].

Depending on the metal present at the active site, hydrogenases can be classified into three classes, namely FeFe hydrogenase, NiFe hydrogenase, and Fe hydrogenase. As may be expected, each of these classes shows different catalytic capacities for H_2 production [52]. NiFe hydrogenase bound to O_2-tolerant membranes tends to catalyze the oxidation of H_2 and produce H_2 under only acidic conditions [53]. By contrast, FeFe hydrogenase produces H_2 in neutral media and at a higher turnover frequency. This is due to the gas access channel of FeFe hydrogenases being significantly shorter than that of NiFe hydrogenases. Unlike NiFe hydrogenases, FeFe hydrogenases are not significantly inhibited by the accumulation of H_2 [54]. However, FeFe hydrogenases are sensitive to reversible O_2 inhibition, with active sites potentially being degraded by light, posing limitations to practical applications [55]. In Fe-hydrogenases, H_2 is activated only in the presence of a substrate (meteniltetrahidro-metanopterina) [56]. Based on the catalytic activity and structure of hydrogenases, a variety of biomolecular catalysts have been synthesized, aiming to reproduce the active sites observed with hydrogenases, often achieving equal or better catalytic efficiency [57–59].

As FeFe hydrogenases are able to produce H_2 easier and under mild conditions, as well as with a higher turnover frequency, these enzymes have been the focus in most of the research conducted related to the bioelectrosynthesis of H_2. FeFe hydrogenases can catalyze the reversible interconversion of protons and electrons into H_2, but generally catalyze the production of H_2 [60]. Depending on their location in the cell, hydrogenases will have different functions in microorganisms. Such an example is periplasmic FeFe hydrogenase from *Desulfovibrio desulfuricans* (*Dd*H), which is involved in the uptake of H_2. Here, protons resulting from the oxidation of H_2 generate a gradient across the membrane, which appear to be coupled with ATP synthesis in the cytoplasm. By contrast, FeFe hydrogenase I from *Clostridium pasteurianum* (*Cp*I) is a cytoplasmic enzyme that accepts electrons from ferredoxin and reduces protons to H_2, to regenerate oxidized ferredoxin [61].

Despite some structural differences between FeFe hydrogenases found in different organisms and their location in the microbial cell, all the studied FeFe hydrogenases share a common architecture comprising several FeS cluster domains linked to the catalytic H cluster domain [62, 63]. This H cluster domain is also common to all FeFe hydrogenases and houses the active site. It consists of a cubane [4Fe–4S] cluster linked to a [2Fe] unit, coordinated to five CN^- and CO diatomic ligands as well as a non-protein dithiolate ligand. The additional FeS cluster domains work toward coupling reversible H_2 oxidation to the reduction or oxidation of external electron donors/acceptors involved in various metabolic processes [61, 64].

Bioelectrosynthesis of H_2 by hydrogenases can be performed by employing the direct electron transfer (DET) mechanism, if the FeS cluster closest to the enzyme surface (distal) is facing the electrode surface. This allows the transfer of electrons between the FeS cluster and the electrode surface [65]. Most electrochemical studies have shown that DET can be achieved by the direct adsorption of small amounts of enzymes on the rough surfaces of carbon electrodes [66, 67]. The roughness of the carbon surface enables a significant amount of enzymes to have a distal cluster close to the electrode surface, facilitating electron transfer [68].

The two main limitations to the use of hydrogenases in practical applications are their sensitivity toward high potential deactivation and oxygen inhibition. To solve this problem, a multilayer polymer bioanode design has been proposed, where the hydrogenase is protected from high potential deactivation and oxygen damage [69]. For instance, NiFe hydrogenase from Desulfovibrio vulgaris Miyazaki F (*DvM*-NiFe) and NiFeSe hydrogenase from Desulfovibrio vulgaris Hildenborough (*DvH-NiFeSe*) were connected to a glassy carbon electrode using a viologen-modified polymer. This polymer has a low redox potential and provided protection against high potential deactivation. The electrode was further covered with a bi-enzymatic protection system against oxygen; the former consisted of oxidase and catalase enzymes trapped in the matrix of the redox-silent hydrophilic polymer (Fig. 8). When this bioanode was

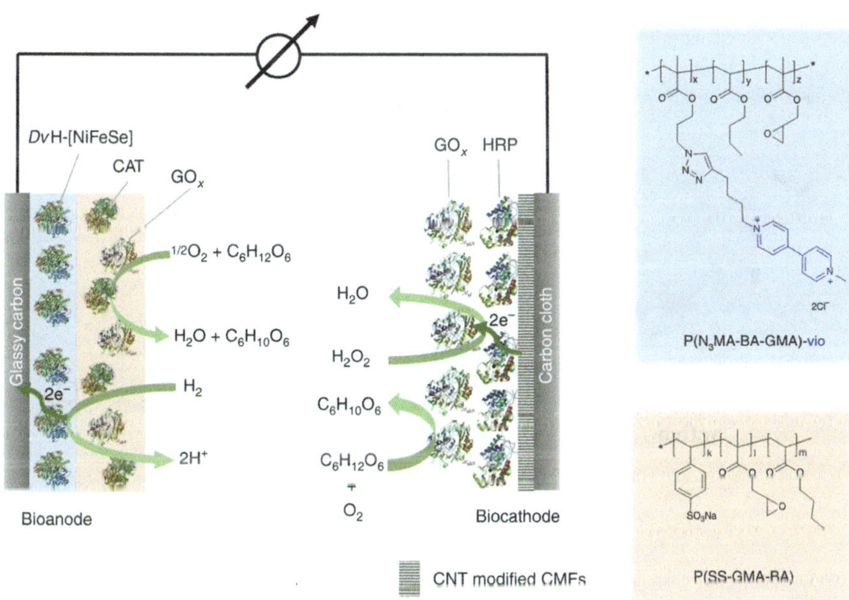

Fig. 8 Schematic representation of the hydrogenase/horseradish peroxidase biofuel cell. Reprinted with permission from [69]. Copyright 2021 American Chemical Society

tested in a H_2/glucose fuel cell, the O_2 removal system, consisting of oxidase/catalase enzymes, was fed with glucose. The glucose also acted as a reagent for the in situ generation of H_2O_2 at the oxidase/horseradish peroxidase-based biocathode. This fuel cell successfully obtained an open circuit reference voltage of 1.15 V, with up to 530 μW cm^{-2} power densities at 0.85 V, even in the presence of O_2 [69].

Some naturally occurring hydrogenases have a mechanism to protect against O_2 attack, where they show a catalytically inactive state (H_{inact}) in the presence of O_2, which then prevents O_2 binding. This H_{inact} state was originally identified in the isolated form of the FeFe hydrogenase (*Dd*H) from *Desulfovibrio desulfuricans*. Although H_{inact} provides protection against O_2 before the first reductive activation of H_{ox} (active state), after gaining catalytic activity, *Dd*H does become sensitive to O_2 again. Through the addition of sulfide under oxidizing conditions, an H_{inact} state for *Dd*H and some other FeFe hydrogenases can also be generated. FeFe hydrogenase from *Clostridium beijerinckii* CbA5H is the only enzyme currently characterized that shows reversible H_{inact}, whereby it can be reduced back to the active state without the addition of sulfide (Fig. 9). The X-ray structure of CbA5H, when exposed to air, revealed that a cysteine residue, preserved in the local environment of the active site (H cluster), directly coordinates with the substrate-binding site, preventing O_2

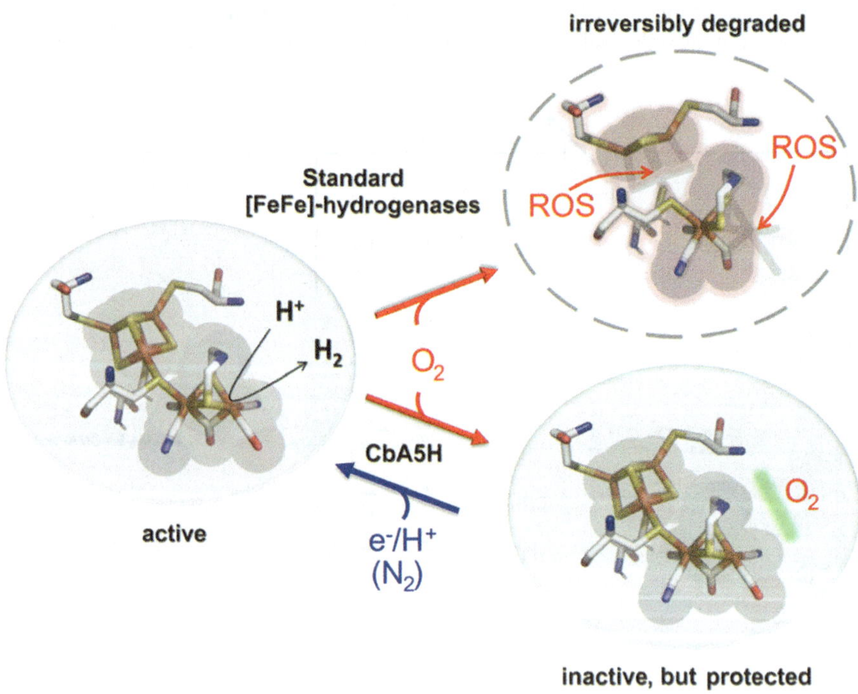

Fig. 9 Schematic illustrating the O_2-resistance of [FeFe]-hydrogenase CbA5H, with ROS being reactive O_2-species resulting from O_2-activation. The green bar indicates an unknown protection mechanism. Reprinted with permission from [70]. Copyright 2021, American Chemical Society

binding [70]. It can therefore be concluded that the discovery of new enzymes and strategies immobilization /bioelectrode design have allowed the use of hydrogenases for efficient production of H_2 without inhibition by O_2.

5 Final Considerations

Significant advances have been made in the field of bioelectrosynthesis of both value-added compounds and fuels in recent years. This is mainly due to the development of new electrode materials and design along with an increased understanding of the catalytic, electron transfer, and inhibition mechanisms. However, challenges in achieving more robust bioelectrodes and improving the electron-transfer rate at the enzyme/electrode interface should be still reached before practical applications in large-scale industrial processes.

References

1. Oliveira, A.R., Mota, C., Mourato, C., Domingos, R.M., Santos, M.F.A., Gesto, D., Guigliarelli, B., Santos-Silva, T., Romão, M.J., Cardoso Pereira, I.A.: Toward the mechanistic understanding of enzymatic CO_2 reduction. ACS Catal. **10**, 3844–3856 (2020). https://doi.org/10.1021/acscatal.0c00086
2. Cadoux, C., Milton, R.D.: Recent enzymatic electrochemistry for reductive reactions. ChemElectroChem **7**, 1974–1986 (2020). https://doi.org/10.1002/celc.202000282
3. Gu, W., Milton, R.D.: Natural and engineered electron transfer of nitrogenase. Chemistry (Easton) **2**, 322–346 (2020). https://doi.org/10.3390/chemistry2020021
4. Foster, S.L., Bakovic, S.I.P., Duda, R.D., Maheshwari, S., Milton, R.D., Minteer, S.D., Janik, M.J., Renner, J.N., Greenlee, L.F.: Catalysts for nitrogen reduction to ammonia. Nat. Catal. **1**, 490–500 (2018). https://doi.org/10.1038/s41929-018-0092-7
5. Jang, J., Jeon, B.W., Kim, Y.H.: Bioelectrochemical conversion of CO_2 to value added product formate using engineered Methylobacterium extorquens. Sci. Rep. **8**, 7211 (2018). https://doi.org/10.1038/s41598-018-23924-z
6. Chen, J.G., Crooks, R.M., Seefeldt, L.C., Bren, K.L., Bullock, R.M., Darensbourg, M.Y., Holland, P.L., Hoffman, B., Janik, M.J., Jones, A.K., Kanatzidis, M.G., King, P., Lancaster, K.M., Lymar, S. V, Pfromm, P., Schneider, W.F., Schrock, R.R.: Beyond fossil fuel-driven nitrogen transformations. Science **360**(80), eaar6611 (2018). https://doi.org/10.1126/science.aar6611
7. Bajracharya, S., Vanbroekhoven, K., Buisman, C.J.N., Strik, D.P.B.T.B., Pant, D.: Bioelectrochemical conversion of CO_2 to chemicals: CO_2 as a next generation feedstock for electricity-driven bioproduction in batch and continuous modes. Faraday Discuss. **202**, 433–449 (2017). https://doi.org/10.1039/C7FD00050B
8. Chen, H., Prater, M.B., Cai, R., Dong, F., Chen, H., Minteer, S.D.: Bioelectrocatalytic conversion from N_2 to chiral amino acids in a H_2/α-keto acid enzymatic fuel cell. J. Am. Chem. Soc. **142**, 4028–4036 (2020). https://doi.org/10.1021/jacs.9b13968
9. Milton, R.D., Minteer, S.D.: Enzymatic bioelectrosynthetic ammonia production: recent electrochemistry of nitrogenase, nitrate reductase, and nitrite reductase. Chempluschemistry **82**, 513–521 (2017). https://doi.org/10.1002/cplu.201600442

10. Martins, M., Pereira, I.A.C., Pita, M., De Lacey, A.L.: Biological production of hydrogen BT—Enzymes for solving humankind's problems: natural and artificial systems in health, agriculture, environment and energy (2021)
11. Holladay, J.D., Hu, J., King, D.L., Wang, Y.: An overview of hydrogen production technologies. Catal. Today. **139**, 244–260 (2009). https://doi.org/10.1016/j.cattod.2008.08.039
12. Armaroli, N., Balzani, V.: The hydrogen issue. ChemSusChem **4**, 21–36 (2011). https://doi.org/10.1002/cssc.201000182
13. Wang, J.-W., Liu, W.-J., Zhong, D.-C., Lu, T.-B.: Nickel complexes as molecular catalysts for water splitting and CO_2 reduction. Coord. Chem. Rev. **378**, 237–261 (2019). https://doi.org/10.1016/j.ccr.2017.12.009
14. Milton, R.D., Abdellaoui, S., Khadka, N., Dean, D.R., Leech, D., Seefeldt, L.C., Minteer, S.D.: Nitrogenase bioelectrocatalysis: heterogeneous ammonia and hydrogen production by MoFe protein. Energy Environ. Sci. **9**, 2550–2554 (2016). https://doi.org/10.1039/C6EE01432A
15. Pereira, A.R., Sedenho, G.C., Souza, J.C.P.D.E., Crespilho, F.N.: Advances in enzyme bioelectrochemistry. An. Acad. Bras. Cienc. **90**, 825–857 (2018)
16. Alissandratos, A., Easton, C.J.: Biocatalysis for the application of CO_2 as a chemical feedstock. Beilstein J. Org. Chem. **11**, 2370–2387 (2015). https://doi.org/10.3762/bjoc.11.259
17. Lubitz, W., Ogata, H., Rüdiger, O., Reijerse, E.: Hydrogenases. Chem. Rev. **114**, 4081–4148 (2014). https://doi.org/10.1021/cr4005814
18. Badalyan, A., Yang, Z.-Y., Seefeldt, L.C.: A voltammetric study of nitrogenase catalysis using electron transfer mediators. ACS Catal. **9**, 1366–1372 (2019). https://doi.org/10.1021/acscatal.8b04290
19. Hu, Y., Ribbe, M.W.: Nitrogenase and homologs. JBIC J. Biol. Inorg. Chem. **20**, 435–445 (2015). https://doi.org/10.1007/s00775-014-1225-3
20. Cai, R., Minteer, S.D.: Nitrogenase bioelectrocatalysis: from understanding electron-transfer mechanisms to energy applications. ACS Energy Lett. **3**, 2736–2742 (2018). https://doi.org/10.1021/acsenergylett.8b01637
21. Seefeldt, L.C., Hoffman, B.M., Dean, D.R.: Mechanism of Mo-dependent nitrogenase. Annu. Rev. Biochem. **78**, 701–722 (2009). https://doi.org/10.1146/annurev.biochem.78.070907.103812
22. Rohde, M., Sippel, D., Trncik, C., Andrade, S.L.A., Einsle, O.: The critical E_4 state of nitrogenase catalysis. Biochemistry **57**, 5497–5504 (2018). https://doi.org/10.1021/acs.biochem.8b00509
23. Neese, F.: The Yandulov/Schrock cycle and the nitrogenase reaction: pathways of nitrogen fixation studied by density functional theory. Angew. Chemie Int. Ed. **45**, 196–199 (2006). https://doi.org/10.1002/anie.200502667
24. Yandulov, D.V., Schrock, R.R.: Catalytic reduction of dinitrogen to ammonia at a single molybdenum center. Science **301**, 76–78 (2003). https://doi.org/10.1126/science.1085326
25. Kästner, J., Blöchl, P.E.: Ammonia production at the FeMo cofactor of nitrogenase: results from density functional theory. J. Am. Chem. Soc. **129**, 2998–3006 (2007). https://doi.org/10.1021/ja068618h
26. Hoffman, B.M., Lukoyanov, D., Dean, D.R., Seefeldt, L.C.: Nitrogenase: a draft mechanism. Acc. Chem. Res. **46**, 587–595 (2013). https://doi.org/10.1021/ar300267m
27. Jasniewski, A.J., Lee, C.C., Ribbe, M.W., Hu, Y.: Reactivity, mechanism, and assembly of the alternative nitrogenases. Chem. Rev. **120**, 5107–5157 (2020). https://doi.org/10.1021/acs.chemrev.9b00704
28. Chen, H., Simoska, O., Lim, K., Grattieri, M., Yuan, M., Dong, F., Lee, Y.S., Beaver, K., Weliwatte, S., Gaffney, E.M., Minteer, S.D.: Fundamentals, applications, and future directions of bioelectrocatalysis. Chem. Rev. **120**, 12903–12993 (2020). https://doi.org/10.1021/acs.chemrev.0c00472
29. Hickey, D.P., Cai, R., Yang, Z.-Y., Grunau, K., Einsle, O., Seefeldt, L.C., Minteer, S.D.: Establishing a thermodynamic landscape for the active site of Mo-dependent nitrogenase. J. Am. Chem. Soc. **141**, 17150–17157 (2019). https://doi.org/10.1021/jacs.9b06546

30. Hickey, D.P., Lim, K., Cai, R., Patterson, A.R., Yuan, M., Sahin, S., Abdellaoui, S., Minteer, S.D.: Pyrene hydrogel for promoting direct bioelectrochemistry: ATP-independent electroenzymatic reduction of N_2. Chem. Sci. **9**, 5172–5177 (2018). https://doi.org/10.1039/C8SC01638K
31. Lee, Y.S., Yuan, M., Cai, R., Lim, K., Minteer, S.D.: Nitrogenase bioelectrocatalysis: ATP-independent ammonia production using a redox polymer/MoFe protein system. ACS Catal. **10**, 6854–6861 (2020). https://doi.org/10.1021/acscatal.0c01397
32. Milton, R.D., Cai, R., Abdellaoui, S., Leech, D., De Lacey, A.L., Pita, M., Minteer, S.D.: Bioelectrochemical Haber–Bosch process: an ammonia-producing H_2/N_2 fuel cell. Angew. Chemie Int. Ed. **56**, 2680–2683 (2017). https://doi.org/10.1002/anie.201612500
33. Maia, L.B., Moura, I., Moura, J.J.G.: Molybdenum and tungsten-containing formate dehydrogenases: aiming to inspire a catalyst for carbon dioxide utilization. Inorganica Chim. Acta. **455**, 350–363 (2017). https://doi.org/10.1016/j.ica.2016.07.010
34. Schlager, S., Dumitru, L.M., Haberbauer, M., Fuchsbauer, A., Neugebauer, H., Hiemetsberger, D., Wagner, A., Portenkirchner, E., Sariciftci, N.S.: Electrochemical reduction of carbon dioxide to methanol by direct injection of electrons into immobilized enzymes on a modified electrode. Chemsuschem **9**, 631–635 (2016). https://doi.org/10.1002/cssc.201501496
35. Vo, T., Purohit, K., Nguyen, C., Biggs, B., Mayoral, S., Haan, J.L.: Formate: an Energy Storage and Transport Bridge between Carbon Dioxide and a Formate Fuel Cell in a Single Device. Chemsuschem **8**, 3853–3858 (2015). https://doi.org/10.1002/cssc.201500958
36. Zhang, L., Ong, J., Liu, J., Li, S.F.Y.: Enzymatic electrosynthesis of formate from CO_2 reduction in a hybrid biofuel cell system. Renew. Energy. **108**, 581–588 (2017). https://doi.org/10.1016/j.renene.2017.03.009
37. Amao, Y., Shuto, N.: Formate dehydrogenase catalyzed CO_2 reduction in a chlorin-e6 sensitized photochemical biofuel cell. J. Porphyr. Phthalocyanines. **19**, 459–464 (2015). https://doi.org/10.1142/S1088424615500406
38. Cordas, C.M., Campaniço, M., Baptista, R., Maia, L.B., Moura, I., Moura, J.J.G.: Direct electrochemical reduction of carbon dioxide by a molybdenum-containing formate dehydrogenase. J. Inorg. Biochem. **196**, 110694 (2019). https://doi.org/10.1016/j.jinorgbio.2019.110694
39. Calzadiaz-Ramirez, L., Calvó-Tusell, C., Stoffel, G.M.M., Lindner, S.N., Osuna, S., Erb, T.J., Garcia-Borràs, M., Bar-Even, A., Acevedo-Rocha, C.G.: In vivo selection for formate dehydrogenases with high efficiency and specificity toward NADP+. ACS Catal. **10**, 7512–7525 (2020). https://doi.org/10.1021/acscatal.0c01487
40. Niks, D., Hille, R.: Chapter eleven—Reductive activation of CO_2 by formate dehydrogenases. In: Armstrong, F. (ed) Methods in Enzymology. pp. 277–295. Academic Press (2018)
41. Guo, Q., Gakhar, L., Wickersham, K., Francis, K., Vardi-Kilshtain, A., Major, D.T., Cheatum, C.M., Kohen, A.: Structural and Kinetic studies of formate dehydrogenase from Candida boidinii. Biochemistry **55**, 2760–2771 (2016). https://doi.org/10.1021/acs.biochem.6b00181
42. Reda, T., Plugge, C.M., Abram, N.J., Hirst, J.: Reversible interconversion of carbon dioxide and formate by an electroactive enzyme. Proc. Natl. Acad. Sci. **105**, 10654 (2008). https://doi.org/10.1073/pnas.0801290105
43. Robinson, W.E., Bassegoda, A., Blaza, J.N., Reisner, E., Hirst, J.: Understanding how the rate of C–H bond cleavage affects formate oxidation catalysis by a Mo-dependent formate dehydrogenase. J. Am. Chem. Soc. **142**, 12226–12236 (2020). https://doi.org/10.1021/jacs.0c03574
44. Robinson, W.E., Bassegoda, A., Reisner, E., Hirst, J.: Oxidation-state-dependent binding properties of the active site in a Mo-containing formate dehydrogenase. J. Am. Chem. Soc. **139**, 9927–9936 (2017). https://doi.org/10.1021/jacs.7b03958
45. Miller, M., Robinson, W., Oliveira, A., Heidary, N., Kornienko, N., Warnan, J., Cardoso Pereira, I., Reisner, E.: Interfacing formate dehydrogenase with metal oxides for reversible electrocatalysis and solar-driven reduction of carbon dioxide. Angew. Chemie. **131** (2019). https://doi.org/10.1002/ange.201814419
46. Kuk, S.K., Gopinath, K., Singh, R.K., Kim, T.-D., Lee, Y., Choi, W.S., Lee, J.-K., Park, C.B.: NADH-free electroenzymatic reduction of CO_2 by conductive hydrogel-conjugated formate dehydrogenase. ACS Catal. **9**, 5584–5589 (2019). https://doi.org/10.1021/acscatal.9b00127

47. Maia, L.B., Fonseca, L., Moura, I., Moura, J.J.G.: Reduction of carbon dioxide by a molybdenum-containing formate dehydrogenase: a Kinetic and mechanistic study. J. Am. Chem. Soc. **138**, 8834–8846 (2016). https://doi.org/10.1021/jacs.6b03941
48. Yuan, M., şahin, S., Cai, R., Abdellaoui, S., Hickey, D., Minteer, S., Milton, R.: Creating a low-potential redox polymer for efficient electroenzymatic CO_2 reduction. Angew. Chemie Int. Ed. **57**, (2018). https://doi.org/10.1002/anie.201803397
49. Min, K., Park, Y.-S., Park, G.W., Lee, J., Moon, M., Ko, C.H., Lee, J.-S.: Elevated conversion of CO_2 to versatile formate by a newly discovered formate dehydrogenase from Rhodobacter aestuarii. Bioresour. Technol. **305**, 123155 (2020). https://doi.org/10.1016/j.biortech.2020.123155
50. Stephenson, M., Stickland, L.H.: Hydrogenlyases: bacterial enzymes liberating molecular hydrogen. Biochem. J. **26**, 712–724 (1932). https://doi.org/10.1042/bj0260712
51. Kim, D.H., Kim, M.S.: Hydrogenases for biological hydrogen production. Bioresour. Technol. **102**, 8423–8431 (2011). https://doi.org/10.1016/j.biortech.2011.02.113
52. Hexter, S.V., Grey, F., Happe, T., Climent, V., Armstrong, F.A.: Electrocatalytic mechanism of reversible hydrogen cycling by enzymes and distinctions between the major classes of hydrogenases. Proc. Natl. Acad. Sci. U. S. A. **109**, 11516–11521 (2012). https://doi.org/10.1073/pnas.1204770109
53. Lu, Y., Koo, J.: O_2 sensitivity and H_2 production activity of hydrogenases—A review. Biotechnol. Bioeng. **116**, 3124–3135 (2019). https://doi.org/10.1002/bit.27136
54. Fourmond, V., Baffert, C., Sybirna, K., Dementin, S., Abou-Hamdan, A., Meynial-Salles, I., Soucaille, P., Bottin, H., Léger, C.: The mechanism of inhibition by H_2 of H_2-evolution by hydrogenases. Chem. Commun. **49**, 6840–6842 (2013). https://doi.org/10.1039/C3CC43297A
55. Fontecilla-Camps, J.C., Volbeda, A., Cavazza, C., Nicolet, Y.: Structure/function relationships of [NiFe]- and [FeFe]-hydrogenases. Chem. Rev. **107**, 4273–4303 (2007). https://doi.org/10.1021/cr050195z
56. Yang, X., Hall, M.B.: Monoiron hydrogenase catalysis: hydrogen activation with the formation of a dihydrogen, Fe−Hδ−···Hδ+−O, Bond and Methenyl-H4MPT+ Triggered Hydride Transfer. J. Am. Chem. Soc. **131**, 10901–10908 (2009). https://doi.org/10.1021/ja902689n
57. Schilter, D., Camara, J.M., Huynh, M.T., Hammes-Schiffer, S., Rauchfuss, T.B.: Hydrogenase enzymes and their synthetic models: the role of metal hydrides. Chem. Rev. **116**, 8693–8749 (2016). https://doi.org/10.1021/acs.chemrev.6b00180
58. Simmons, T.R., Berggren, G., Bacchi, M., Fontecave, M., Artero, V.: Mimicking hydrogenases: from biomimetics to artificial enzymes. Coord. Chem. Rev. **270–271**, 127–150 (2014). https://doi.org/10.1016/j.ccr.2013.12.018
59. Gloaguen, F., Rauchfuss, T.B.: Small molecule mimics of hydrogenases: hydrides and redox. Chem. Soc. Rev. **38**, 100–108 (2009). https://doi.org/10.1039/B801796B
60. Tard, C., Pickett, C.J.: Structural and functional analogues of the active sites of the [Fe]-, [NiFe]-, and [FeFe]-hydrogenases. Chem. Rev. **109**, 2245–2274 (2009). https://doi.org/10.1021/cr800542q
61. Mulder, D.W., Shepard, E.M., Meuser, J.E., Joshi, N., King, P.W., Posewitz, M.C., Broderick, J.B., Peters, J.W.: Insights into [FeFe]-hydrogenase structure, mechanism, and maturation. Structure **19**, 1038–1052 (2011). https://doi.org/10.1016/j.str.2011.06.008
62. Mulder, D.W., Boyd, E.S., Sarma, R., Lange, R.K., Endrizzi, J.A., Broderick, J.B., Peters, J.W.: Stepwise [FeFe]-hydrogenase H-cluster assembly revealed in the structure of HydA(DeltaEFG). Nature **465**, 248–251 (2010). https://doi.org/10.1038/nature08993
63. Ogata, H., Kellers, P., Lubitz, W.: The crystal structure of the [NiFe] hydrogenase from the photosynthetic bacterium Allochromatium vinosum: characterization of the oxidized enzyme (Ni-A state). J. Mol. Biol. **402**, 428–444 (2010). https://doi.org/10.1016/j.jmb.2010.07.041
64. Czech, I., Silakov, A., Lubitz, W., Happe, T.: The [FeFe]-hydrogenase maturase HydF from Clostridium acetobutylicum contains a CO and CN^- ligated iron cofactor. FEBS Lett. **584**, 638–642 (2010). https://doi.org/10.1016/j.febslet.2009.12.016
65. Gentil, S., Che Mansor, S.M., Jamet, H., Cosnier, S., Cavazza, C., Le Goff, A.: Oriented immobilization of [NiFeSe] hydrogenases on covalently and noncovalently functionalized carbon

nanotubes for H_2/air enzymatic fuel cells. ACS Catal. **8**, 3957–3964 (2018). https://doi.org/10.1021/acscatal.8b00708

66. Quinson, J., Hidalgo, R., Ash, P.A., Dillon, F., Grobert, N., Vincent, K.A.: Comparison of carbon materials as electrodes for enzyme electrocatalysis: hydrogenase as a case study. Faraday Discuss. **172**, 473–496 (2014). https://doi.org/10.1039/C4FD00058G
67. Alonso-Lomillo, M.A., Rüdiger, O., Maroto-Valiente, A., Velez, M., Rodríguez-Ramos, I., Muñoz, F.J., Fernández, V.M., De Lacey, A.L.: Hydrogenase-coated carbon nanotubes for efficient H_2 oxidation. Nano Lett. **7**, 1603–1608 (2007). https://doi.org/10.1021/nl070519u
68. Baffert, C., Sybirna, K., Ezanno, P., Lautier, T., Hajj, V., Meynial-Salles, I., Soucaille, P., Bottin, H., Léger, C.: Covalent attachment of FeFe hydrogenases to carbon electrodes for direct electron transfer. Anal. Chem. **84**, 7999–8005 (2012). https://doi.org/10.1021/ac301812s
69. Ruff, A., Szczesny, J., Marković, N., Conzuelo, F., Zacarias, S., Pereira, I.A.C., Lubitz, W., Schuhmann, W.: A fully protected hydrogenase/polymer-based bioanode for high-performance hydrogen/glucose biofuel cells. Nat. Commun. **9**, 3675 (2018). https://doi.org/10.1038/s41467-018-06106-3
70. Winkler, M., Duan, J., Rutz, A., Felbek, C., Scholtysek, L., Lampret, O., Jaenecke, J., Apfel, U.-P., Gilardi, G., Valetti, F., Fourmond, V., Hofmann, E., Léger, C., Happe, T.: A safety cap protects hydrogenase from oxygen attack. Nat. Commun. **12**, 756 (2021). https://doi.org/10.1038/s41467-020-20861-2

Progress in Enzyme-Based Biofuel Cells

Graziela C. Sedenho

Abstract Oxidoreductases are highly specific and efficient biocatalysts, that can be applied on electrochemical systems for green energy conversion, such as biofuel cells. Thanks to the redox enzyme biocatalysts, biofuel cells can operate under mild conditions of pH, temperature, and pressure. In this context, the present chapter addresses recent aspects of application of enzymatic bioelectrocatalysis in biofuel cells, including advances, issues, and challenges for real-world applications.

1 Introduction

Biofuel cells (BFCs) have gained great attention in the last decades, as they are sustainable energy conversion systems and employ renewable catalysts, the redox enzymes (oxidoreductases) [1]. Redox enzymes (oxidoreductases) are natural biocatalysts present in all living organisms on earth and are intrinsically related to the metabolic energy obtention. Unlike the traditional inorganic electrocatalysts used in energy conversion systems, redox enzymes are renewable, able to operate under mild conditions of pH, temperature, and pressure, and show wide substrate scope and high activity and selectivity [1–3]. These features become redox enzymes attractive biocatalysts for applications on several systems and devices, including bioelectrochemical energy conversion systems, as BFCs [1].

Oxidoreductases have been employed as electrocatalysts in both electrodes of BFCs: in the anodes to catalyze the electron transfer from the fuel to the electrode surface, and in the cathodes to catalyze the electron transfer from the electrode surface to the oxidant agent, usually molecular oxygen (Fig. 1) [1]. Remarkably, an interesting application field of enzymatic BFCs is as energy sources of implantable biomedical devices and self-powered biosensors. Among the oxidoreductases, the most common classes of enzymes employed in BFCs as bioelectrocatalysts are dehydrogenases, hydrogenases, and oxidases.

G. C. Sedenho (✉)
São Carlos Institute of Chemistry, University of São Paulo (USP), Avenida Trabalhador São-carlense, 400, São Carlos, SP 13560-970, Brazil
e-mail: graziela.cs@usp.br

F. N. Crespilho (ed.), *Advances in Bioelectrochemistry Volume 4*,
https://doi.org/10.1007/978-3-030-99662-8_4

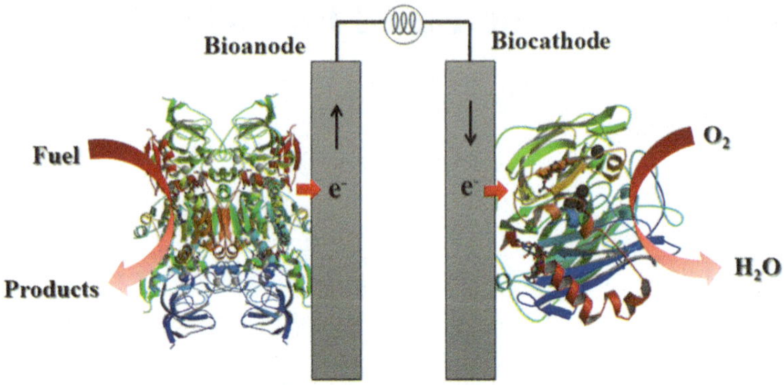

Fig. 1 Representative scheme of an enzyme-based BFC

Dehydrogenases are able to catalyze the reversible hydride transfer reactions between their coenzymes or cofactors, such as NAD, FAD, and quinone derivatives, and a diversity of organic compounds, including alcohols, aldehydes, and sugars. The most used dehydrogenases in the bioanodes of BFCs are alcohol dehydrogenase (ADH), glucose dehydrogenase (GDh), lactate dehydrogenase, fructose dehydrogenase, formate dehydrogenase, and malate dehydrogenase, which receive their name according to the compounds (fuels) that are able to oxidize [4]. Hydrogenases catalyze the reversible oxidation of molecular hydrogen into protons and electrons, and have been efficiently employed as biocatalysts in bioanodes of H_2/O_2 BFCs to replace the traditional platinum-based catalysts because of their high turnover frequency and small overpotential [5]. Differently, oxidases have been extensively employed as biocatalyst in both anodes and cathodes of BFCs. The most widely enzyme used in bioanodes is glucose oxidase (GOD), which oxidates glucose into gluconolactone [6], whereas multicopper oxidases (MCOs), especially laccase and bilirubin oxidase (BOD), have been widely used in biocathodes for catalytic reduction of molecular oxygen into water [7]. The most exploited reaction in cathodes of (bio)fuel cells is the molecular oxygen redox reaction (ORR), due to the wide availability of this oxidant agent in the nature. In this context, the present chapter, divided by the types of enzymes, addresses recent and innovative aspects of enzyme-based BFCs, including advances, issues, and challenges.

2 Alcohol Dehydrogenase-Based Bioanodes

ADH from *Saccharomyces cerevisiae* has been extensively used as anode catalyst of innumerous ethanol/O_2 enzymatic fuel cells, as it catalyzes the oxidation of ethanol to acetaldehyde with the interconversion of NAD^+ to NADH, as shown in Fig. 2a [1]. In bioelectrochemical systems involving NAD-dependent enzymes, the electrons

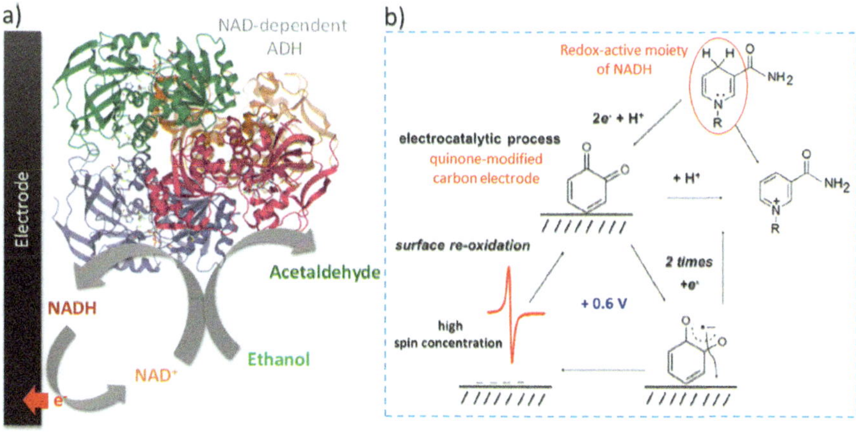

Fig. 2 **a** Scheme of the bioelectrochemical oxidation of ethanol by ADH (PDB 4W6Z) and electro-regeneration of NAD^+. **b** Simplified mechanism of electron transfer between NADH and quinone-functionalized carbon electrodes. Adapted with permission from *Ali MA, Hassan A, Sedenho GC, Gonçalves R V, Cardoso DR, Crespilho FN (2019) Operando Electron Paramagnetic Resonance for Elucidating the Electron Transfer Mechanism of Coenzymes. J. Phys. Chem. C 123:16,058 –16,064.* Copyright 2019 American Chemical Society

generated by the fuel oxidation in the active site of the enzyme are shuttled to the electrode surface through reduced coenzyme NADH. Therefore, the bioelectrocatalytic performance of the bioanode is dependent on the efficient oxidation of NADH to NAD^+ on the electrode surface (Fig. 2a). In addition, the maintenance of the biocatalytic cycle also depends on the regeneration of the oxidized state of the coenzyme. Therefore, the efficient electrooxidation of NADH to NAD^+ is of particular interest to researchers in the construction of high-performance bioanodes for BFCs.

In this context, many efforts have been made to achieve the electrochemical oxidation of NADH at low overpotentials, to obtain BFCs with satisfactory cell voltage. The use of redox polymer mediators is a common strategy to facilitate the fast electron transfer between the NADH and the carbon-based electrode surface, not only in bioelectrochemical systems involving ADH, but also other NAD-dependent enzymes, such as glucose dehydrogenase. The redox mediator should be carefully chosen, as it should provide low overpotentials for the electrochemical NAD^+ regeneration [8]. Examples of redox mediators successfully used for this purpose are: poly(methylene green) [9, 10], quinone-based polymer [11], viologen-based polymers [12], and organometallic ruthenium compound [13]. Alternatively, the introduction of oxygenated-functional groups and defects on the basal planes of the carbon-based electrode surfaces by chemical pretreatment has demonstrated to be a promising way to provide efficient NADH oxidation in ADH-based bioanodes. The quinone-like structures on the carbon electrode surface formed by the chemical oxidation enhance the electrooxidation of NADH [14, 15]. Electron paramagnetic resonance spectroscopic measurements under electrochemical control have shown quinone groups on carbon electrode surface act as electrocatalysts toward the oxidation of NADH to NAD^+ (Fig. 2b), differently of redox mediators [15].

Ethanol is a very attractive fuel because is safe and renewable, shows very low toxicity, has high energy density, and is easily and widely produced from fermentation process. These features have contributed to the interest of several researchers in developing ethanol BFCs. Recently, it was demonstrated the possibility to integrate in situ ethanol bioproduction and oxidation into a single system for electricity production. Glucose was used as a substrate for *Saccharomyces cerevisiae* produces ethanol through fermentation, while ADH was used to catalyze the oxidation of the in situ-generated ethanol to acetaldehyde, delivering electrons to the electrode without redox mediators [16]. This work shows the concept that microorganisms and enzymes can function cooperatively in a BFC, simultaneously taking advantage of the microbial robustness and substrate flexibility, and the fast catalytic and electron transfer kinetics of redox enzymes, without the need to use sophisticated procedures [16]. Other advance in ethanol BFCs is the use of enzyme cascades for complete ethanol oxidation for improving the power output. ADH was combined with aldehyde dehydrogenase in the bioanode for oxidation of ethanol to CO_2, collecting up to 12 electrons per ethanol molecule [17]. However, despite the several improvements, the main challenge of this approach is still the collection of all electrons released during the complete fuel oxidation by the multistep cascade, to obtain maximum faradaic efficiency and energy output.

3 Hydrogenase-Based Bioanodes

Since 2006 purified hydrogenases have been widely exploited as biocatalyst in anodes of H_2/O_2 enzymatic fuel cells due to their attractive features toward the oxidation of dihydrogen [18]. Although hydrogenases are the most active biocatalyst for H_2/H^+ interconversion, the practical application of these redox enzymes in anodes of H_2/O_2 BFCs faces some issues, such as deactivation at high potentials and by reaction with O_2 during the BFC operation. The catalytic site of the [NiFe] is inactivated by the formation of two oxidized states, Ni–A and Ni–B. Even the O_2-tolerant [NiFe] hydrogenases undergo inactivation by O_2, forming a Ni–B state, however their Ni–B active state is faster reestablished than the standard [NiFe] hydrogenases [19]. At higher potentials, in direct electron transfer systems, the hydrogenases are inactivated by the oxidation of the active site, leading to the formation of the inactive Ni–B state. In this context, several strategies have been emerged to protect both [NiFe] and [FeFe] hydrogenases from O_2 and high-potential deactivation. Some interesting strategies involve the blocking of O_2 to the catalytic site or the conversion of O_2 into a harmless product [20].

Some studies have reported the use of viologen derivatives to prevent O_2 to reach the active layer of enzyme on the electrode surface and the high-potential inactivation [21–23]. Thick redox viologen-based hydrogel films are able to react with O_2 producing water, protecting the hydrogenase from O_2 damage and remaining the biocatalytic current quietly unaffected [21]. Additionally, the viologen hydrogel [21] protects the enzyme to experience high potentials, as the active site will be submitted only to the formal potential of the viologen, independently of the potential applied on the electrode [21]. A similar effect was verified with viologen derivative covalently attached to the electrode surface, and allows electrons to tunnel directly from the enzyme to the electrode surface, even in the presence of the viologen layer [22]. More recently, thinner films based on viologen-modified dendrimers have shown to prevent hydrogenase damage by O_2, and do not require high load of enzyme nor cause mass transport limitations, as observed in thicker films of redox hydrogel [23]. Theoretical calculations have recently shown the effect of enzymatic redox polymers films on the penetration of O_2, and suggested thinner films (ca. 6 μm) combine excellent catalyst use, current, and long-lasting protection [24].

The advances in hydrogenase protection against high potential and O_2 deactivation added to design engineering have significantly contributed to the development of H_2/O_2 BFCs in recent years. For example, [NiFe] hydrogenase has been successfully used as biocatalyst in dual-gas breathing hydrogen/air membrane-free BFC, reaching maximum power output of 3.6 mW cm^{-2} at 0.7 V, under anode-limiting conditions. This was possible developing a bioanode consisting of a two-layer system with a viologen-modified polymer-based adhesion layer and an active redox polymer/hydrogenase top layer, as shown in Fig. 3 [25]. Also, several studies have shown the use of viologen-based redox polymer to enable the use of high O_2-sensitive [FeFe] hydrogenase in gas-diffusion membrane-free H_2/O_2 BFC with outstanding power densities and cell voltages close to the thermodynamic limit [26, 27].

4 Glucose Oxidase-Based Bioanodes

GOD has been extensively studied and applied as a biocatalyst in anode of glucose BFCs due to its great ability to catalyze glucose oxidation. The development of high-performance glucose BFCs is of great interest of bioelectrochemical and clean energy fields, because glucose is a readily abundant and clean fuel, as well as, is widely available in biological fluids and tissues of living organisms. Because of that, some of the most promising and attractive applications of glucose BFCs are on implantable devices and on self-powered biosensors [28, 29]. Such devices have been developed for biomedical purposes, such to monitor health conditions in real time, to be integrated with other devices harvesting energy to them, such pacemakers, artificial organs, electrical stimulators, and wireless transmission systems to deliver health parameters to an external device, or even for environment monitoring [30–38]. In this sense, glucose BFCs have been implanted in rats, insects, crustaceans, mollusks, and fruits.

The successful application of implanted glucose BFCs into living beings requires miniaturized design, biocompatibility, as well as high energy density and stable systems, as all other BFCs. It is evident that the application of GOD as biocatalyst in glucose BFCs is only possible due to the development of strategies to overcome

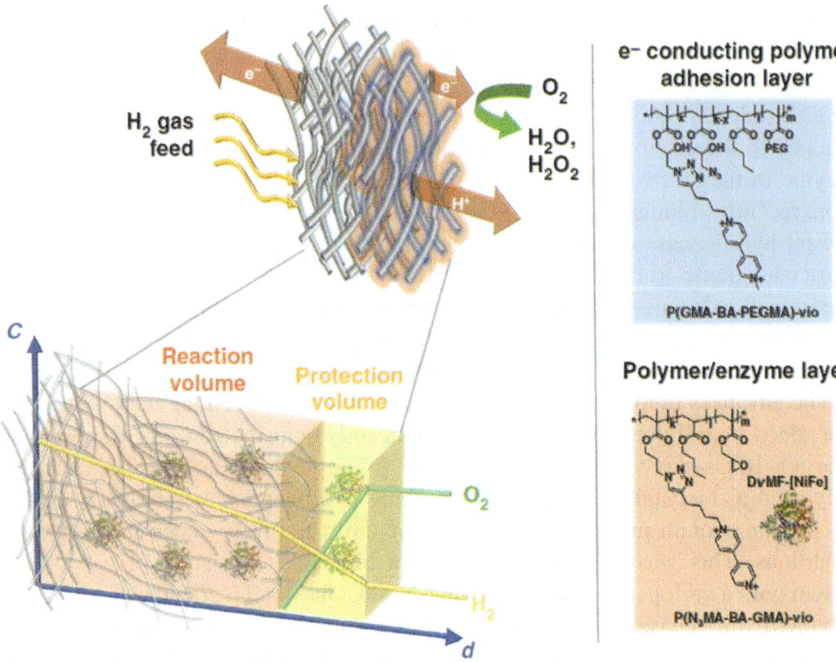

Fig. 3 Scheme of a gas-breathing redox polymer-based hydrogenase bioanode for protection against O_2 and high-potential deactivation. Reprinted from Szczesny et al. [25]. *A gas breathing hydrogen/air biofuel cell comprising a redox polymer/hydrogenase-based bioanode. Nat. Commun. 9: 4715.* Copyright 2018, Springer Nature

the O_2 competitive effect and the slow electron transfer rate that would drastically harm the performance of the GOD bioanode, as discussed in a previous chapter. In this sense, many efforts have been devoted to develop biocompatible, miniaturized, membraneless glucose/O_2 BFCs. A BFC without a membrane separating the anodic and cathodic chambers is always desirable, because this component generally increases ohmic resistance and the cost of the system, and becomes the miniaturization of the device complicated [28]. Additional care must be taken with the use of GOD to avoid production of toxic H_2O_2 in implantable and wearable BFCs, as O_2 is always present in the atmosphere and in biological environments. In this sense, the combination of GOD with catalase [30, 31, 39] and horseradish peroxidase [40] immobilized on anode surface has shown to be effective to circumvent this issue in implantable devices. Similar to catalase, polydopamine coating the electrode has been used to catalyze the H_2O_2 decomposition [41].

The miniaturization of glucose BFCs for implantable or wearable applications is an important aspect, as large-sized devices make the implantation process difficult or even impossible, and may require too invasive surgery or cause excessive discomfort to the living being. In this sense, pellet-type bioelectrodes with 0.5 cm^2 and 1 mm thickness were developed for implantation in the abdomen of a rabbit (Fig. 4a) [31];

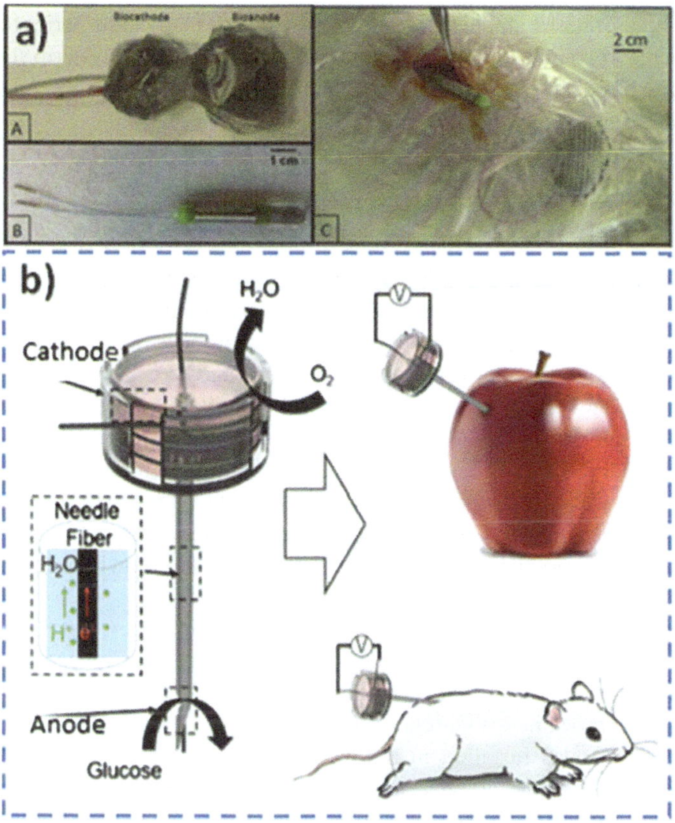

Fig. 4 **a** Bioelectrodes (top-left) and tele-transmission system (bottom-left) that were implanted in the abdomen of a rabbit (right). Reprinted from *Electrochim. Acta, 269, Ichi-Ribault S El, Alcaraz J, Boucher F, Boutaud B, Dalmolin R, Boutonnat J, Cinquin P, Zebda A, Martin DK, Remote wireless control of an enzymatic biofuel cell implanted in a rabbit for 2 months, 360–366*, https://doi.org/10.1016/j.electacta.2018.02.156. Copyright (2020), with permission from Elsevier. **b** BFC assembled into a needle and inserted into an apple and a mouse. Adapted from *Biosens. Bioelectron., 165, Yin S, Liu X, Kobayashi Y, Nishina Y, Nakagawa R, Yanai R, Kimura K, Miyake T, A needle-type biofuel cell using enzyme/mediator/carbon nanotube composite fibers for wearable electronics, 112,287*, Copyright (2020), with permission from Elsevier

a needly-type glucose-O_2 BFC was assembled using the needle bioanode and O_2-diffusion biocathode to be partially inserted into abdominal cavity and heart of mice and into fruits (Fig. 4b) [42]; a catheter-type BFC using flexible carbon fibers as electrodes was implanted into the jugular vein of rat [35].

Several advances has been made in glucose BFCs, such as overcoming the O_2 competitive effect, improving the electron transfer rate of enzymes, the power output, and the biocompatibility, and miniaturizing systems, however, achieving considerable long-term stability is still a challenge, as enzymes usually lose their catalytic activity over time. Some biomedical implantable devices, especially pacemakers and artificial organs, require stable glucose BFCs over years to avoid surgeries for replacements. Therefore, many advances must be performed to reach long-term stability and enable those practical applications in humans.

5 Multicopper Oxidases-based biocathodes

MCOs, particularly laccase and BOD, are able to reduce O_2 to water at small overpotentials and without generation of any toxic intermediate, because of that they have been extensively employed as biocatalysts in cathodes of BFCs, [7]. The ORR, widely used in cathodes of both traditional fuel cells and BFCs, shows two main limitations: (1) the low solubility of O_2 in aqueous electrolytes, approximately 0.25 mmol L^{-1} at room temperature and atmospheric pressure; (2) the slow diffusion of O_2 into aqueous electrolytes, which leads the reaction to be limited by the mass transport of O_2 to the electrode [43]. Usually, the slow diffusion of O_2 toward the electrode represents the bottleneck of the ORR. These two facts are the most significant causes of inefficiency in common BFCs, therefore several studies in recent decades have aimed to overcome this limitation.

To overcome these limitations related to gaseous substances, gas diffusion electrodes (GDE) have emerged. This type of electrode has been successfully used in fuel cells with a proton exchange membrane [44, 45] and, more recently, in the development of biocathodes with laccase and BOD for O_2 reduction [46–49]. GDEs are porous electrodes installed in a way that one side faces the electrolyte and the other side faces the gas phase. For that, they should be mainly composed of a hydrophobic layer (or gas diffusion layer) and a catalytic layer and provide a large three-phase boundary surface. The hydrophobic layer is usually formed by a mixture of carbon materials, such as carbon nanotubes and carbon nanoparticles, and binders for constructing a mesoporous structure with high porosity, gas permeability, and electrical conductivity, and that should prevent the electrolyte solution leakage. The catalytic layer contains the immobilized redox enzyme and, if necessary, redox polymers or hydrogels can be used in this layer [43]. Sometimes, a basement can be used to support the catalytic and the hydrophobic layers, for example, carbon cloths, felts, and papers.

In this sense, combining the advantages of GDE and an efficient biocatalyst, such as BOD, properly immobilized on the catalytic layer surface, high-performance biocathodes for ORR can be achieved to be used in BFCs. Electrode engineering and studies of enzyme immobilization have significantly contributed to BOD-based GDEs with higher current densities and satisfactory operational stability, which are desirable for practical application in BFCs. The immobilization of BOD on carbon nanotubes [50, 51] or carbon black [52, 53] has provided increased gas permeability and biocatalytic current densities higher than 10 mA cm^{-2} toward ORR in absence of redox mediator and in pH 5, value close to the optimal pH for the enzyme activity. In conditions similar to physiological environment, i.e., pH 7.0–7.2, O_2 reduction current densities higher than 1 mA cm^{-2} have been reported with BOD immobilized on carbon nanotubes [42, 54–56], modified carbon cloth [57], and carbon nanoparticles [58]. Concerning operational stability, the incorporation of BOD into a gel matrix based on Nafion and glutaraldehyde have been reported to create a suitable microenvironment for long-term stability under continuous operation combined with satisfactory catalytic current without redox mediator, as shown in Fig. 5 [58].

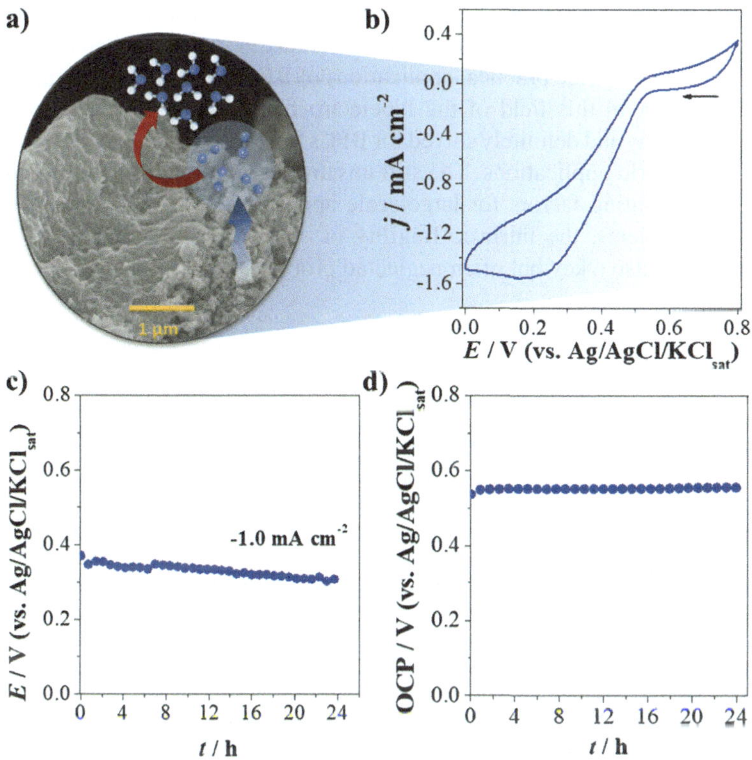

Fig. 5 **a** Scheme of ORR on a gas diffusion biocathode based on BOD incorporate in a biogel matrix. **b** Electrochemical performance of the biocathode and stability under, **c** operation and **d** open circuit potential (OCP) at pH 7.2, 25 °C, under atmospheric condition. Reprinted from *J. Power Sources, 2021, 482, Sedenho GC, Hassan A, Macedo LJA, Crespilho FN, Stabilization of bilirubin oxidase in a biogel matrix for high-performance gas diffusion electrodes, 229,035*. Copyright (2021), with permission from Elsevier

6 Final Considerations

In recent decades, research in BFCs has made great progress, mainly involving the understanding of biocatalytic and electron transfer mechanisms, as well as development of more attractive enzyme immobilization methods for overcoming or minimizing inactivation processes and improving the operational stability of bioelectrodes, which impact the practical application of BFCs. In addition, electrode engineering and cell designing have contributed to achieve systems with enhanced performance and enabled some practical applications of BFCs in laboratory scale. Despite several advances in this field of the bioelectrochemistry, many crucial problems should be urgently and definitely solved for BFCs moving from fundamental research to some real-world applications. The still unsatisfactory efficiency and stability are the common limiting factors for large-scale application of BFCs. Apart from the mentioned problems, the intrinsic fragility of enzyme catalyst out of its natural environment is also a key but often neglected problem.

References

1. Luz, R.A.S., Pereira, A.R., de Souza, J.C.P., Sales, F.C.P.F., Crespilho, F.N.: Enzyme biofuel cells: thermodynamics, kinetics and challenges in applicability. ChemElectroChem **1**, 1751–1777 (2014). https://doi.org/10.1002/celc.201402141
2. Chen, H., Simoska, O., Lim, K., Grattieri, M., Yuan, M., Dong, F., Lee, Y.S., Beaver, K., Weliwatte, S., Gaffney, E.M., Minteer, S.D.: Fundamentals, applications, and future directions of bioelectrocatalysis. Chem. Rev. **120**, 12903–12993 (2020). https://doi.org/10.1021/acs.chemrev.0c00472
3. Chen, H., Dong, F., Minteer, S.D.: The progress and outlook of bioelectrocatalysis for the production of chemicals, fuels and materials. Nat. Catal. **3**, 225–244 (2020). https://doi.org/10.1038/s41929-019-0408-2
4. Zhang, J.L., Wang, Y.H., Huang, K., Huang, K.J., Jiang, H., Wang, X.M.: Enzyme-based biofuel cells for biosensors and in vivo power supply. Nano Energy **84**, 105853 (2021). https://doi.org/10.1016/j.nanoen.2021.105853
5. Lojou, E.: Hydrogenases as catalysts for fuel cells: strategies for efficient immobilization at electrode interfaces. Electrochim Acta **56**, 10385–10397 (2011). https://doi.org/10.1016/j.electacta.2011.03.002
6. Cosnier, S., Gross, A.J., Giroud, F., Holzinger, M.: Beyond the hype surrounding biofuel cells: what's the future of enzymatic fuel cells? Curr. Opin. Electrochem. **12**, 148–155 (2018). https://doi.org/10.1016/j.coelec.2018.06.006
7. Mano, N., de Poulpiquet, A.: O_2 Reduction in enzymatic biofuel cells. Chem. Rev. **118**, 2392–2468 (2017). https://doi.org/10.1021/acs.chemrev.7b00220
8. Lee, Y.S., Gerulskis, R., Minteer, S.D.: Advances in electrochemical cofactor regeneration: enzymatic and non-enzymatic approaches. Curr. Opin. Biotechnol. **73**, 14–21 (2022). https://doi.org/10.1016/j.copbio.2021.06.013
9. Aquino Neto, S., Milton, R.D., Hickey, D.P., De Andrade, A.R., Minteer, S.D.: Membraneless enzymatic ethanol/O_2 fuel cell: transitioning from an air-breathing Pt-based cathode to a bilirubin oxidase-based biocathode. J. Power Sour. **324**, 208–214 (2016). https://doi.org/10.1016/j.jpowsour.2016.05.073

10. Neto, S.A., Minteer, S.D., de Andrade, A.R.: Developing ethanol bioanodes using a hydrophobically modified linearpolyethylenimine hydrogel for immobilizing an enzyme cascade. J. Electroanal Chem. **812**, 153–158 (2018)
11. Abdellaoui, S., Milton, R.D., Quah, T., Minteer, S.D.: NAD-dependent dehydrogenase bioelectrocatalysis: the ability of a naphthoquinone redox polymer to regenerate NAD. Chem. Commun. **52**, 1147–1150 (2016). https://doi.org/10.1039/c5cc09161f
12. Quah, T., Milton, R.D., Abdellaoui, S., Minteer, S.D.: Bioelectrocatalytic NAD+/NADH interconversion: transformation of an enzymatic fuel cell into an enzymatic redox flow battery. Chem. Commun. **53**, 8411 (2017)
13. Lalaoui, N., Means, N., Walgama, C., Le Goff, A., Holzinger, M., Krishnan, S., Cosnier, S.: Enzymatic versus electrocatalytic oxidation of NADH at carbon-nanotube electrodes modified with glucose dehydrogenases: application in a bucky-paper-based glucose enzymatic fuel cell. ChemElectroChem **3**, 2058–2062 (2016). https://doi.org/10.1002/celc.201600545
14. Pereira, A.R., Souza, J.C.P., Gonçalves, A.D., Pagnoncelli, K.C., Crespilho, F.N.: Bioelectrooxidation of ethanol using NAD-dependent alcohol dehydrogenase on oxidized flexible carbon fiber arrays. J. Braz. Chem. Soc. **28**, 1698–1707 (2017)
15. Ali, M.A., Hassan, A., Sedenho, G.C., Gonc, R.V., Cardoso, D.R., Crespilho, F.N.: Operando electron paramagnetic resonance for elucidating the electron transfer mechanism of coenzymes. J. Phys. Chem. C **123**, 16058–16064 (2019). https://doi.org/10.1021/acs.jpcc.9b01160
16. Pagnoncelli, K.C., Pereira, A.R., Sedenho, G.C., Bertaglia, T., Crespilho, F.N.: Ethanol generation, oxidation and energy production in a cooperative bioelectrochemical system. Bioelectrochemistry **122**, 11–25 (2018). https://doi.org/10.1016/j.bioelechem.2018.02.007
17. Franco, J.H., Neto, S.A., Hickey, D.P., Minteer, S.D., de Andrade, A.R.: Hybrid catalyst cascade architecture enhancement for complete ethanol electrochemical oxidation. Biosens Bioelectron **121**, 281–286 (2018). https://doi.org/10.1016/j.bios.2018.09.011
18. Vincent, K.A., Cracknell, J.A., Clark, J.R., Ludwig, M., Lenz, O., Friedrich, B., Armstrong, F.A.: Electricity from low-level H_2 in still air—An ultimate test for an oxygen tolerant hydrogenase. Chem. Commun. 5033–5035 (2006). https://doi.org/10.1039/b614272a
19. Hamdan, A.A., Burlat, B., Gutiérrez-Sanz, O., Liebgott, P.P., Baffert, C., De Lacey, A.L., Rousset, M., Guigliarelli, B., Léger, C., Dementin, S.: O_2-independent formation of the inactive states of NiFe hydrogenase. Nat Chem Biol **9**, 15–17 (2013). https://doi.org/10.1038/nchembio.1110
20. Lu, Y., Koo, J.: O_2 sensitivity and H_2 production activity of hydrogenases-a review. Biotechnol. Bioeng. **116**, 3124–3135 (2019). https://doi.org/10.1002/bit.27136
21. Plumeré, N., Rüdiger, O., Oughli, A.A., Williams, R., Vivekananthan, J., Pöller, S., Schuhmann, W., Lubitz, W.: A redox hydrogel protects hydrogenase from high-potential deactivation and oxygen damage. Nat. Chem. **6**, 822–827 (2014). https://doi.org/10.1038/nchem.2022
22. Oughli, A.A., Vélez, M., Birrell, J.A., Schuhmann, W., Lubitz, W., Plumeré, N., Rüdiger, O.: Viologen-modified electrodes for protection of hydrogenases from high potential inactivation while performing H_2 oxidation at low overpotential. Dalt. Trans. **47**, 10685–10691 (2018). https://doi.org/10.1039/c8dt00955d
23. Li, H., Buesen, D., Dementin, S., Léger, C., Fourmond, V., Plumeré, N.: Complete protection of O_2-sensitive catalysts in thin films. J. Am. Chem. Soc. **141**, 16734–16742 (2019). https://doi.org/10.1021/jacs.9b06790
24. Fourmond, V., Léger, C.: Theoretical understanding of the penetration of O_2 in enzymatic redox polymer films: the case of unidirectional catalysis and irreversible inactivation in a film of arbitrary thickness. ChemElectroChem **8**, 2607–2615 (2021). https://doi.org/10.1002/celc.202100586
25. Szczesny, J., Markovi, N., Conzuelo, F., Zacarias, S., Pereira, I.A.C., Plumeré, N., Schuhmann, W., Ruff, A., Lubitz, W.: A gas breathing hydrogen/air biofuel cell comprising a redox polymer/hydrogenase-based bioanode. Nat. Commun. **9**, 4715 (2018). https://doi.org/10.1038/s41467-018-07137-6
26. Szczesny, J., Birrell, J.A., Conzuelo, F., Lubitz, W., Ruff, A., Schuhmann, W.: Redox-polymer-based high-current-density gas-diffusion H_2-oxidation bioanode using [FeFe] hydrogenase

from desulfovibrio desulfuricans in a membrane-free biofuel cell. Angew. Chemie—Int. Ed. **59**, 16506–16510 (2020). https://doi.org/10.1002/anie.202006824
27. Hardt, S., Stapf, S., Filmon, D.T., Birrell, J.A., Rüdiger, O., Fourmond, V., Léger, C., Plumeré, N.: Reversible H_2 oxidation and evolution by hydrogenase embedded in a redox polymer film. Nat. Catal. **4**, 251–258 (2021). https://doi.org/10.1038/s41929-021-00586-1
28. Sedenho, G.C., Pereira, A.R., Pagnoncelli, K.C., De, S.J.C.P., Crespilho, F.N.: Implantable enzyme-based biofuel cells. In: Wandelt, K. (ed.) Encyclopedia of Interfacial Chemistry: Surface Science and Electrochemistry, pp. 248–260. Elsevier Inc., Amsterdam (2018)
29. Wang, L., Wu, X., Su, B.S.Q., Song, R., Zhang, J.-R., Zhu, J.-J.: Enzymatic biofuel cell: opportunities and intrinsic challenges in futuristic applications. Adv. Energy Sustain. Res. **2**, 2100031 (2021). https://doi.org/10.1002/aesr.202100031
30. Zebda, A., Cosnier, S., Alcaraz, J.-P., Holzinger, M., Goff, A., Le Gondran, C., Boucher, F., Giroud, F., Gorgy, K., Lamraoui, H., Cinquin, P.: Single glucose biofuel cells implanted in rats power electronic devices. Sci. Rep. **3**, 1516 (2013). https://doi.org/10.1038/srep01516
31. El Ichi-Ribault, S., Alcaraz, J., Boucher, F., Boutaud, B., Dalmolin, R., Boutonnat, J., Cinquin, P., Zebda, A., Martin, D.K.: Remote wireless control of an enzymatic biofuel cell implanted in a rabbit for 2 months. Electrochim. Acta **269**, 360–366 (2018). https://doi.org/10.1016/j.electacta.2018.02.156
32. Lopez, F., Zerria, S., Ruff, A., Schuhmann, W.: An O_2 tolerant polymer/glucose oxidase based bioanode as basis for a self-powered glucose sensor. Electroanalysis **30**, 1311–1318 (2018). https://doi.org/10.1002/elan.201700785
33. Lee, D., Jeong, S.H., Yun, S., Kim, S., Sung, J., Seo, J., Son, S., Kim, J.T., Susanti, L., Jeong, Y., Park, S., Seo, K., Kim, S.J., Chung, T.D.: Totally implantable enzymatic biofuel cell and brain stimulator operating in bird through wireless communication. Biosens. Bioelectron. **171**, 112746 (2021). https://doi.org/10.1016/j.bios.2020.112746
34. Mano, N., Mao, F., Heller, A.: Characteristics of a miniature compartment-less glucose-O_2 biofuel cell and its operation in a living plant. J. Am. Chem. Soc. **125**, 6588–6594 (2003). https://doi.org/10.1021/ja0346328
35. Sales, F.C.P.F., Iost, R.M., Martins, M.V.A., Almeida, M.C., Crespilho, F.N.: An intravenous implantable glucose/dioxygen biofuel cell with modified flexible carbon fiber electrodes. Lab. Chip. **13**, 468–474 (2013). https://doi.org/10.1039/C2LC41007A
36. Rasmussen, M., Ritzmann, R.E., Lee, I., Pollack, A.J., Scherson, D.: An implantable biofuel cell for a live insect. J. Am. Chem. Soc. **134**, 1458–1460 (2012)
37. MacVittie, K., Halámek, J., Halámková, L., Southcott, M., Jemison, W.D., Lobeld, R., Katz, E., Halamek, J., Halámková, L., Southcott, M., Jemison, W.D., Lobel, R., Katz, E.: From "Cyborg" lobsters to a pacemaker powered by implantable biofuel cells. Energy Environ. Sci. **6**, 81–86 (2013). https://doi.org/10.1039/c2ee23209j
38. Halámková, L., Halámek, J., Bocharova, V., Szczupak, A., Alfonta, L., Katz, E.: Implanted biofuel cell operating in a living snail. J. Am. Chem. Soc. **134**, 5040–5043 (2012)
39. Cinquin, P., Gondran, C., Giroud, F., Mazabrard, S., Pellissier, A., Boucher, F.F., Alcaraz, J.P., Gorgy, K., Lenouvel, F., Mathé, S., Porcu, P., Cosnier, S.: A glucose biofuel cell implanted in rats. PLoS ONE **5**, 1–7 (2010). https://doi.org/10.1371/journal.pone.0010476
40. Yimamumaimaiti, T., Lu, X., Zhang, J.R., Wang, L., Zhu, J.J.: Efficient blood-toleration enzymatic biofuel cell via in situ protection of an enzyme catalyst. ACS Appl. Mater. Interfaces **12**, 41429–41436 (2020). https://doi.org/10.1021/acsami.0c11186
41. Chen, H., Ru, X., Wang, H., Liu, P., Li, G., Cao, Y., Bai, Z., Yang, L.: Construction of a cascade catalyst of nanocoupled living red blood cells for implantable biofuel cell. ACS Appl. Mater. Interfaces **13**, 28010–28016 (2021). https://doi.org/10.1021/acsami.1c01479
42. Yin, S., Liu, X., Kobayashi, Y., Nishina, Y., Nakagawa, R., Yanai, R., Kimura, K., Miyake, T.: A needle-type biofuel cell using enzyme/mediator/carbon nanotube composite fibers for wearable electronics. Biosens. Bioelectron. **165**, 112287 (2020). https://doi.org/10.1016/j.bios.2020.112287
43. Horst, A.E.W., Mangold, K.M., Holtmann, D.: Application of gas diffusion electrodes in bioelectrochemical syntheses and energy conversion. Biotechnol. Bioeng. **113**, 260–267 (2016). https://doi.org/10.1002/bit.25698

44. Paganin, V.A., Ticianelli, E.A., Gonzalez, E.R.: Development and electrochemical studies of gas diffusion electrodes for polymer electrolyte fuel cells. J. Appl. Electrochem. **26**, 297–304 (1996). https://doi.org/10.1007/BF00242099
45. Colmati, F., Paganin, V.A., Gonzalez, E.R.: Effect of operational parameters of mini-direct methanol fuel cells operating at ambient temperature. J. Appl. Electrochem. **36**, 17–23 (2006). https://doi.org/10.1007/s10800-005-9019-5
46. Rincón, R.A., Lau, C., Luckarift, H.R., Garcia, K.E., Adkins, E., Johnson, G.R., Atanassov, P.: Enzymatic fuel cells: integrating flow-through anode and air-breathing cathode into a membrane-less biofuel cell design. Biosens. Bioelectron. **27**, 132–136 (2011). https://doi.org/10.1016/j.bios.2011.06.029
47. Gupta, G., Lau, C., Rajendran, V., Colon, F., Branch, B., Ivnitski, D., Atanassov, P.: Direct electron transfer catalyzed by bilirubin oxidase for air breathing gas-diffusion electrodes. Electrochem. Commun. **13**, 247–249 (2011). https://doi.org/10.1016/j.elecom.2010.12.024
48. Gupta, G., Lau, C., Branch, B., Rajendran, V., Ivnitski, D., Atanassov, P.: Direct bioelectrocatalysis by multi-copper oxidases: gas-diffusion laccase-catalyzed cathodes for biofuel cells. Electrochim. Acta **56**, 10767–10771 (2011). https://doi.org/10.1016/j.electacta.2011.01.089
49. Santoro, C., Babanova, S., Erable, B., Schuler, A., Atanassov, P.: Bilirubin oxidase based enzymatic air-breathing cathode: operation under pristine and contaminated conditions. Bioelectrochemistry **108**, 1–7 (2016). https://doi.org/10.1016/j.bioelechem.2015.10.005
50. So, K., Onizuka, M., Komukai, T., Kitazumi, Y., Shirai, O., Kano, K.: Binder/surfactant-free biocathode with bilirubin oxidase for gas-diffusion-type system. Electrochem. Commun. **66**, 58–61 (2016). https://doi.org/10.1016/j.elecom.2016.02.023
51. So, K., Ozawa, H., Onizuka, M., Komukai, T., Kitazumi, Y., Shiraia, O., Kano, K.: Highly permeable gas diffusion electrodes with hollow carbon nanotubes for bilirubin oxidase-catalyzed dioxygen reduction. Electrochim. Acta **246**, 794–799 (2017). https://doi.org/10.1016/j.electacta.2017.06.117
52. So, K., Kitazumi, Y., Shirai, O., Nishikawa, K., Higuchi, Y., Kano, K.: Direct electron transfer-type dual gas diffusion H_2/O_2 biofuel cells. J. Mater. Chem. A **4**, 8742–8749 (2016). https://doi.org/10.1039/c6ta02654k
53. Xia, H., So, K., Kitazumi, Y., Shirai, O., Nishikawa, K.: Dual gas-diffusion membrane- and mediatorless dihydrogen/air-breathing biofuel cell operating at room temperature. J Power Sour. **335**, 105–112 (2016). https://doi.org/10.1016/j.jpowsour.2016.10.030
54. De, P.A., Haddad, R., Le, G.A., Holzinger, M., Mermoux, M., Infossi, P., Mano, N., Lojou, E., Cosnier, S.: A membraneless air-breathing hydrogen biofuel cell based on direct wiring of thermostable enzymes on carbon nanotube electrodes. Chem. Commun. **51**, 7447–7450 (2015). https://doi.org/10.1039/C5CC02166A
55. Lalaoui, N., Holzinger, M., Le, G.A., Cosnier, S.: Diazonium functionalisation of carbon nanotubes for specific orientation of multicopper oxidases: controlling electron entry points and oxygen diffusion to the enzyme. Chem. Eur. J. **22**, 10494–10500 (2016). https://doi.org/10.1002/chem.201601377
56. Yin, S., Jin, Z., Miyake, T.: Wearable high-powered biofuel cells using enzyme/carbon nanotube composite fibers on textile cloth. Biosens. Bioelectron. **141**, 111471 (2019). https://doi.org/10.1016/j.bios.2019.111471
57. Conzuelo, F., Szczesny, J., Markovic, N., Ruff, A., Schuhmann, W., Horna, D.: An air breathing carbon cloth-based screen-printed electrode for applications in enzymatic biofuel cells. Electroanalysis **31**, 217–221 (2019). https://doi.org/10.1002/elan.201800462
58. Sedenho, G.C., Hassan, A., Macedo, L.J.A., Crespilho, F.N.: Stabilization of bilirubin oxidase in biogel matrix for high-performance bioelectrochemical oxygen reduction reaction. J. Power Sour. **482**, 229035 (2021). https://doi.org/10.1016/j.jpowsour.2020.229035

Bioinspired Batteries: Using Nature-Inspired Materials in Greener and Safer Energy Storage Technologies

Thiago Bertaglia, Luana Cristina Italiano Faria, José Eduardo dos Santos Clarindo, and Frank N. Crespilho

Abstract The widespread use of fossil-based energy sources has led to serious environmental problems, including the greenhouse effect. There is currently an intense search underway to find sustainable ways to generate and store energy and replace fossil fuels. Researchers are actively developing alternatives that are based on natural, biological structures or processes. This worldwide process effort is known as bioinspiration. In this context, quinone and flavin derivatives, which resemble the active molecules in bioenergetic cycles, have been successfully applied in energy-storage devices such as redox flow batteries. These compounds undergo fast reactions, are composed of Earth-abundant elements, and are inexpensive; thus, they can accelerate the advancement of reliable and green energy-storage technology. Furthermore, the need for less invasive and more comfortable approaches to medical care has led to the development of implantable, ingestible, and wearable electronic devices. However, current power sources do not meet the safety and structural requirements of these devices. Bioinspired materials, such as biopolymers, have been extensively explored to produce flexible and smart hydrogels, which can replace liquid electrolytes, and allow the construction of user-friendly, flexible batteries. Hence, recent advancements related to the use of bioinspired materials in battery design and fabrication are discussed in this chapter.

1 Introduction

There is currently an urgent need for novel energy-storage devices (ESDs) in a diverse range of technologies. ESDs are required at every size scale, both large and small. At the larger end of the scale, new approaches are needed to address concerns with intermittent renewable energy sources that include solar and wind power. Intermittency can lead to energy shortages that cause human and economic losses [1, 2]. At the smaller end of the scale, the next wave of healthcare-related

T. Bertaglia · L. C. Italiano Faria · J. E. dos Santos Clarindo · F. N. Crespilho (✉)
University of São Paulo, São Carlos Institute of Chemistry, Avenida trabalhador São-carlense, 400, São Carlos, São Paulo, Brazil
e-mail: frankcrespilho@usp.br

F. N. Crespilho (ed.), *Advances in Bioelectrochemistry Volume 4*,
https://doi.org/10.1007/978-3-030-99662-8_5

electronics will obtain physiological data in ways that are minimally invasive and convenient to the user [3]. These aims cannot be achieved using the current rigid power sources, which makes it imperative to develop wearable and flexible ESDs [4, 5]. Thus, researchers are devising strategies to overcome these challenges as well as promote sustainable development on a global basis.

Over billions of years, nature has developed complex processes and hierarchical materials to address the needs of living organisms [6]. Bioenergetic transformations reach their pinnacle in cellular respiration and photosynthesis pathways. These complex pathways use organic molecules, including quinones and flavins, to capture and store energy from the oxidation of small organic molecules and sunlight, respectively [7–9]. Both of these magnificent processes provide a rich treasure-trove for investigators in energy-storage-related fields seeking ideas and alternatives for designing new technologies. At the structural level, nature has countless examples of biopolymers, such as gelatin, chitin, and alginate, which are interesting alternatives for the production of biocompatible, biodegradable, and cheap hydrogels [10]. These soft materials possess flexibility, can achieve high conductivity, and are less prone to leakage, thus meeting the safety and structural requirements of the latest electronic devices [4, 11]. This chapter will cover the recent bioinspired approaches used in the energy-research field. The application of bioinspired molecules in redox flow batteries is reviewed. Finally, the use of natural hydrogels to produce a new generation of aqueous batteries is also discussed.

2 Redox Flow Battery

Redox flow batteries (RFBs) are electrochemical systems capable of converting electrical energy into chemical energy that is stored in tanks outside the electrochemical cell. The full assembly is composed of an electrochemical cell, storage tanks, and pumps to circulate the electrolyte. Figure 1a shows the components of the electrochemical cell, which include endplates, current collectors, graphite flow fields, gaskets, electrodes, and a membrane. Cell function is initiated by pumping the electrolyte from the storage tanks, which brings the active compounds into contact with the electrodes surfaces, where the electrochemical reactions occur (Fig. 1b). The gasket prevents electrolyte leakage and the membrane prevents electrolyte mixing [12].

RFB systems may be classified according to the nature of the active compound, solvent, and physical state of the electrodes. The active compounds employed can be inorganic and organic, for example, in vanadium- and quinone-based RFBs, respectively. The solvent used to carry the active compounds to the electrode surface is either aqueous or non-aqueous, and both types of solvent can be applied in organic and inorganic RFBs. The active compounds can be in one or two physical states: dissolved in the electrolyte or loaded on the electrode surface. In a hybrid flow battery (HFB), one of the compounds is on the electrode [13, 14]. Figures 1b, c show the two different approaches.

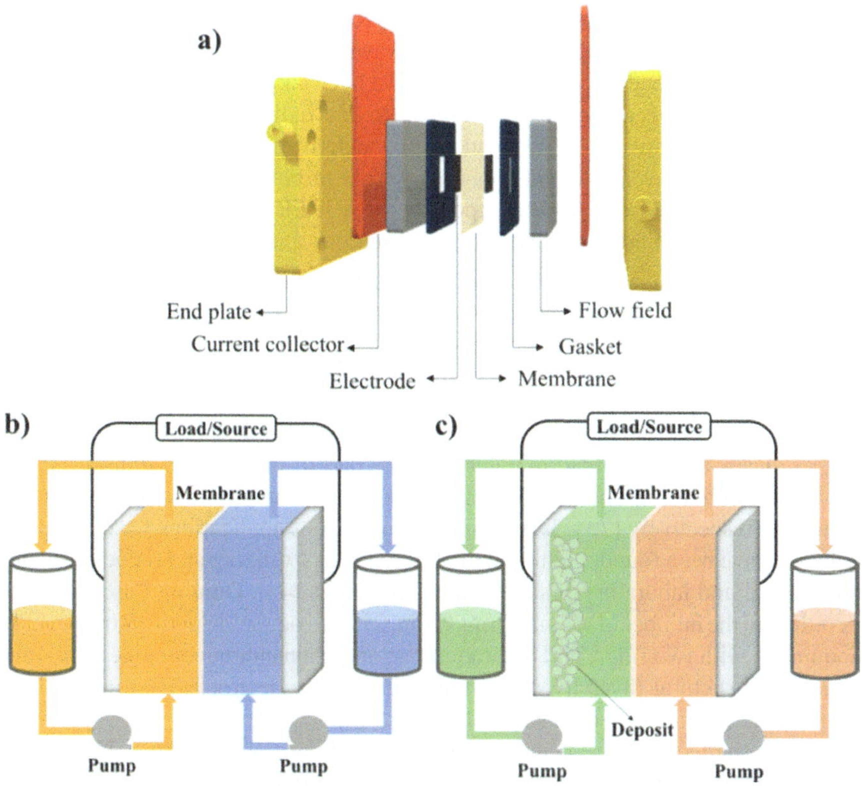

Fig. 1 **a** Illustration of RFB hardware with cell stack components. **b** Schematic representation of a redox flow battery. **c** Schematic representation of a hybrid flow battery. A material is electrodeposited on the electrode when charging the battery

RFBs have a flexible modular design, moderate maintenance costs, excellent scalability, and a long life cycle [15], which makes them a promising solution for the intermittency issues of natural energy sources, such as wind, solar, and tides. Vanadium RFBs (VRFBs) were the first to be explored commercially [16]. This technology uses active vanadium compounds in highly acidic electrolytes (3–4 mol L^{-1} H_2SO_4) that are present on both sides of the battery, with vanadium in all four redox states [17]. Despite their high reversibility and stability, vanadium salts have a high cost, are toxic, can precipitate at both high and low temperatures [18], and have reduced solubility at high acid concentrations [19]. These drawbacks hinder the widespread implementation of the VRFB technology.

Some alternatives to vanadium compounds have been extensively studied. Aqueous organic RFBs (AORFBs) use soluble organic compounds that undergo reversible electrochemical reactions. This type of battery has received much attention during the last two decades. Organic molecules are composed of Earth-abundant elements, may be safe to use, have a potentially low cost [20–23], and can be conveniently changed by inserting donor and withdrawn groups [24, 25]. Some examples of organic/organometallic compounds that have been employed are quinone derivatives [26–33], viologen radicals [34–38], nitroxide radicals (4-HO-TEMPO) [34, 38–40], flavin derivatives [41, 42], and ferrocene derivatives [36, 37]. Among these compounds, quinones and flavins play a key role in biological energy-conversion systems and are promising active compounds for AORFB production. The following section discusses the latest advancements in quinone- and flavin-based AORFBs.

2.1 *Quinone-Based Aqueous Organic Redox Flow Battery*

Quinones are one of the most important and well-studied examples of organic redox couples. They are widespread in nature and are part of the electron transport chain involved in cellular respiration and photosynthesis, for example, ubiquinone and plastoquinone, respectively [7, 43]. Additionally, numerous applications for this type of molecule have been found, including tinctures, and as oxidizing and reducing agents in industrial and laboratory-scale chemical syntheses [43]. Quinones are classified according to the number of aromatic rings (Fig. 2): benzoquinone with one ring, naphthoquinone with two rings, and anthraquinone and phenanthrenequinone with three rings. The structural diversity of quinones allows for electrochemical adaptation, enabling the construction of sustainable ESDs [44].

Quinone molecules have been employed in RFB development since 2009, when Xu et al. [45] reported an HFB using a chloranil-carbon black composite as the cathode and electrodeposited cadmium as the anode. About a year later, the same research group reported an HFB using 1,2-benzoquinone-3,5-disulfonic acid (BQDS) as the positive active compound and a lead plate as the negative active material [46]. These seminal works presented the possibility of building flow batteries using organic molecules, both on the electrode surface and in solution, and created this new research field. Currently, there is a large demand for highly soluble quinone derivatives that can enhance the battery volumetric capacity and power density [47].

The stability of BQDS in acidic media has been extensively studied by Yang et al. [27]. To this end, the authors developed an all-organic aqueous RFB using BQDS as the cathode and anthraquinone-2-sulfonic acid (AQS) as the anode, with both dissolved in 1 mol L^{-1} H_2SO_4 electrolyte. They found that due to the fact that BQDS has sulfonic acid groups in its structure, there is stabilization by intramolecular hydrogen bonding, because benzoquinone has a solubility of 0.1 mol L^{-1}, while BQDS, which has two sulfonic groups, has a solubility of 1.7 mol L^{-1}. This cell showed a capacity-retention rate of 90% at 10 mA cm^{-2} for 12 cycles. In 2016, new studies were conducted using BQDS and 9,10-anthraquinone-2,7-disulfonic acid (AQDS) [48]. The authors employed the NMR technique and proved that BQDS exhibits several chemical and electrochemical transformations during battery cycling and undergoes a Michael reaction. This study demonstrates the necessity of AQDS

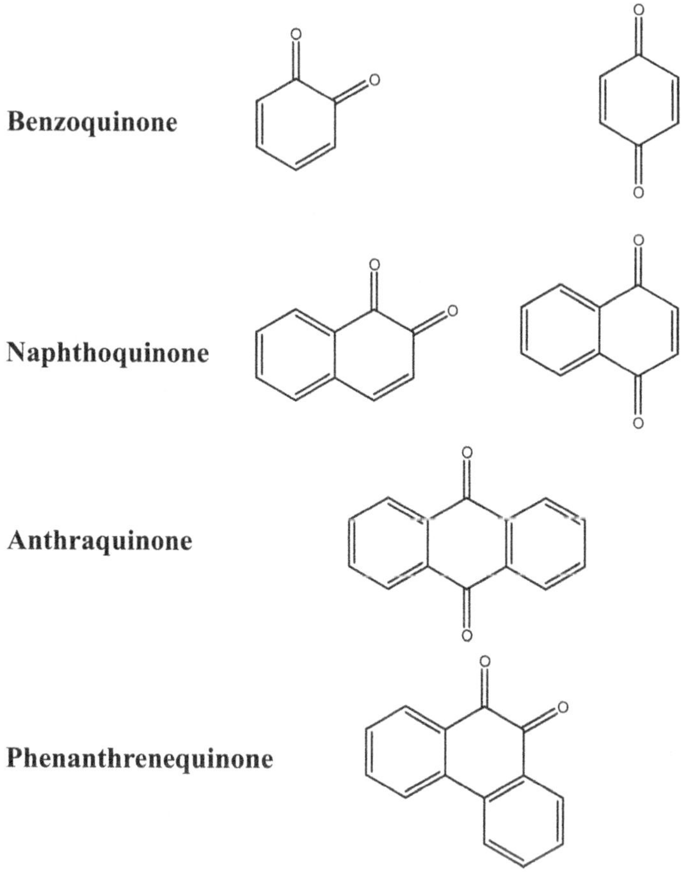

Fig. 2 Different types of quinones and their structures

concentrations that are three times higher than that of BQDS. This battery achieved a coulombic efficiency (CE) of 100% over the course of 100 cycles.

Huskinson et al. [20] reported a high power-density battery that couples the AQDS and Br_2/HBr redox pair under acidic conditions. Cycling studies at 0.2 A cm^{-2} at 50% state of charge (SOC) resulted in reproducible cycles with a current efficiency close to 95%. At a current density of 0.5 A cm^{-2} and a cut-off voltage of 0 V and 1.5 V, the discharge capacity retention was 99%. Furthermore, the polarization and power curves for SOCs ranging from 10 to 90% display an increase in the open-circuit voltage (OCV) and peak power densities from 0.69 V to 0.92 V, and 0.246 W cm^{-2} to 0.600 W cm^{-2}, respectively.

Lin et al. [23] reported an adaptation to an alkaline battery, where bromine was replaced with the non-toxic $K_4[Fe(CN)_6]$ complex, and anthraquinone was functionalized with hydroxyl groups. This approach expanded the battery voltage. The cell was assembled by coupling 2,6-dihydroxyanthraquinone (2,6-DHAQ) and

$K_4[Fe(CN)_6]$ in the negative and positive chambers, respectively. The Pourbaix diagram depicted that 2,6-DHAQ undergoes two-proton two-electron reaction until it reaches pH 12. Above this pH, there is no potential changing since the reduction species were produced in their fully deprotonated form. The charged battery showed an OCV of 1.2 V at 50% SOC and a peak galvanic power density exceeding 0.4 W cm^{-2} in a power curve. Galvanostatic charge/discharge cycles were conducted with a current density of 0.1 A cm^{-2} for 100 cycles, resulting in current and energy efficiencies of 99% and 84%, respectively. In addition, the battery exhibited a capacity fade of 0.1% per cycle over 100 cycles.

Because of the increasing interest in quinones for RFB applications, a theoretical study covering approximately 50 anthraquinone derivatives was carried out by Bachman et al. [25]. They performed density functional theory calculations to determine electrochemical properties, including the substitution effect for electron acceptor and withdrawing groups, redox windows, and free solvation energy. The study showed that substitution by electron donor groups improves the anthraquinone reduction window, with adequate oxidative stability; this is useful for designing new active redox compounds.

Recently, Lee et al. [26] reported an alkaline RFB that uses a mixture of naphthoquinones. The negolyte comprised of a mixture of a sodium salt of 1,2-naphthoquinone-4-sulfonic acid (NQ-S) and 2-hydroxy-1,4-naphthoquinone (Lawsone). $K_3[Fe(CN)_6]$, was used as the posolyte. This mixture of naphthoquinones showed a higher solubility in KOH (1.26 mol L^{-1} in 1 mol L^{-1} KOH) than in individual studies of 0.42 mol L^{-1} and 0.83 mol L^{-1} in 1 mol L^{-1} KOH for Lawsone and NQ-S, respectively. This increase is because NQ-S releases the hydrophilic sulfite ($-SO_3^{2-}$) group during the transformation, and it can form polar-polar interactions with organic species and KOH electrolyte. This battery, with 0.6 mol L^{-1} NQ-SO (mixture of NQ-S and Lawsone) and 0.4 mol L^{-1} $K_4[Fe(CN)_6]$ has a voltage of 1.01 V, with a discharge capacity of 22 Ah L^{-1} at 70% SOC. When the NQ-SO concentration was increased to 1.2 mol L^{-1}, the discharge capacity increased to 40.3 Ah L^{-1} at 83% SOC.

Two new synthetic anthraquinones based on 9,10-dihydroanthracene were developed for use on the negative side of an RFB. 3,3′-(9,10-anthraquinone-diyl)bis(3-methylbutanoic acid) (DPivOHAQ) and 4,4′-(9,10-anthraquinone-diyl)dibutanoic acid (DBAQ) were each paired with $K_4[Fe(CN)_6]$ at pH 12 [49]. DBAQ had a solubility of 1 mol L^{-1}, and DPivOHAQ had a solubility of 0.74 mol L^{-1}. The full cell exhibited a capacity fade rate of 0.0084% per day with DBAQ, and 0.014% per day with DPivOHAQ (Fig. 3a). Furthermore, when DPivOHAQ was exposed to air, and the pH was increased to 14, the cell had a capacity fade rate of 0.0018% per day. This equates to a record low-capacity fade rate of 0.66% per year (Fig. 3a). This change in capacity fade rate with increasing pH was attributed to anthrone formation because an increase in the hydroxide concentration can suppress the formation of anthrone, beyond its exposure to air, and can convert back to anthraquinone. Figures 3b, c show the charge–discharge voltage curves with 0.5 mol L^{-1} DPivOHAQ-$K_4[Fe(CN)_6]$, and indicate a steep profile followed by a short horizontal segment. It was possible to

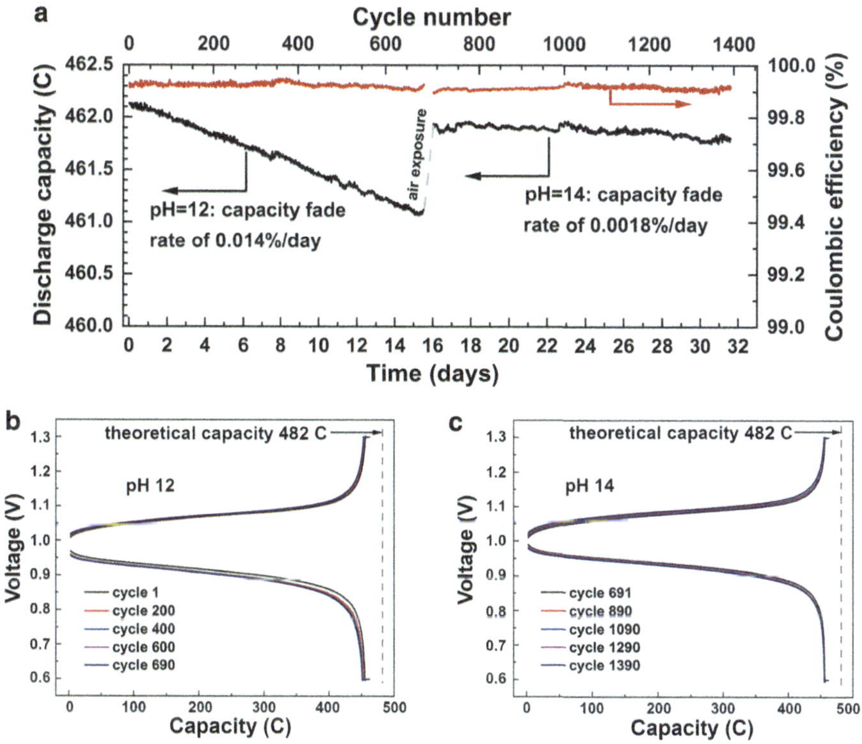

Fig. 3 Cycling performance of DPivOHAQ at pH 12 and 14. **a** CE and discharge capacity versus time and cycle number for a full cell. After approximately 16 days of cycling, DPivOHAQ was exposed to air and the pH of all electrolytes was adjusted to 14 before cycling for another 16 days. **b** Charge–discharge voltage profile of DPivOHAQ from selected cycles at pH 12 in Fig. 3a. **c** Charge–discharge voltage profile of DPivOHAQ from selected cycles at pH 14 in Fig. 3a. Reprinted from Chem, v.6, Wu, M. et al., Extremely Stable Anthraquinone Negolytes Synthesized from Common Precursors, p. 1432–1442, Copyright (2021), with permission from Elsevier

conclude that the electrolyte has approximately 4.5% inactive material, because the horizontal segment ends at 95.5% of the theoretical capacity.

In situ electrosynthesis was conducted by oxidizing a water-soluble anthracene in anthraquinone without the use of oxidants or catalysts, producing the anolyte DPivOHAQ and $K_3[Fe(CN)_6]$ as a catholyte [50]. This is a promising approach because it does not use hazardous materials or require further product purification, which lowers the cost and environmental impact. A charge–discharge test for a single cycle resulted in an OCV equal to 1.0 V, and a capacity of 84 C. Then, 2271 cycles were performed over 33.2 days, with a fade rate of 0.014% per day and CE of 99.53%. The low fade rate was due to the chemical stability of the molecular structure. Therefore, the two branched methyl groups on the anthraquinone core increased the stability of the solubilizing chain. Furthermore, there is a C–C covalent bonding between the anthraquinone core and the functionalizing chains that is more robust at a high pH and elevated temperature.

The influence of the structure of carbon-based electrodes on the electrochemical kinetics of quinones was studied by Sedenho et al. [51]. Glassy carbon (GC), highly ordered pyrolytic graphite (HOPG), and high-edge-density graphite (HEDGE) were chosen as electrode materials because of the different degrees of defects in their structures. Among the three materials, HEDGE presented the highest degree of edge defects, followed by GC, which presented fewer defects, and HOPG, which had almost no defects. Furthermore, the authors employed four quinone derivatives, namely frog quinone (FQ), crab quinone (CQ), BQDS, and Alizarin Red S (ARS), which can be used as a cathode, or in symmetric batteries.

Figure 4 shows the electrochemical behavior of the four quinones studied. All quinones displayed reversible cyclic voltammograms (CV) using the HEDGE electrode, which does not occur with the HOPG and GC electrodes. Furthermore, the CV for HEDGE suggests an electrochemical process that was controlled by adsorbed species on the electrode surface. For instance, the CV of CQ using the HOPG electrode shows a broad oxidation peak and a major reduction peak, with a peak separation (ΔE_p) of 570 mV, which means there is a quasi-reversible reaction. With the GC electrode, the CV for CQ shows a broad oxidation and reduction peak, with a smaller ΔE_p (330 mV). Conversely, the CV of CQ on the HEDGE electrode has well-defined oxidation and reduction peaks, with an ΔE_p of 30 mV, which indicates a reversible reaction controlled by adsorbed species on the electrode surface. Additionally, the CV for ARS shows remarkable differences. GC and HOPG electrodes exhibit similar electrochemical behavior, with a quasi-reversible redox pair at lower potential (approximately −0.20 V) attributed to the 9,10-quinone moiety. Also, the reactions of the 3,4-dihydroxyl groups occur only during the oxidation wave and are considered irreversible. In contrast, the CV of ARS using the HEDGE electrode suggests reversible electrochemical processes at both low and high potentials, with ΔE_p values of 21 and 11 mV, respectively. These results demonstrate the effects of the electrode structure on the electrochemical behavior of the quinones [51].

To understand the electrochemical behavior of HEDGE, its structure was evaluated using spectroscopic measurements and theoretical calculation. Raman measurements showed that HEDGE presents a highly-defective structure, because the I_D/I_G ratio of this electrode was 0.45, which is much higher than the I_D/I_G ratio for HOPG (0.09). Additionally, X-ray photoelectron spectroscopy indicated the presence of oxygenated groups on the HEDGE surface in the form of hydroxyl, ether, and carbonyl functional groups. This structure was further simulated through molecular dynamics, and the role of the defects in the quinone-electrode surface interaction was evaluated. The results show that quinone molecules interact with HEDGE via van der Waals forces (Fig. 5), mainly between the carbon atoms of the carbonyl moieties and the heteroatoms present in the chemical structures of the quinones, such as nitrogen (FQ and CQ) and sulfur (ARS and BQDS). Therefore, the authors concluded that the highly-defective structure of HEDGE and the functional groups present on its surface stabilize the quinone derivatives, which favors their electrochemical kinetics [51].

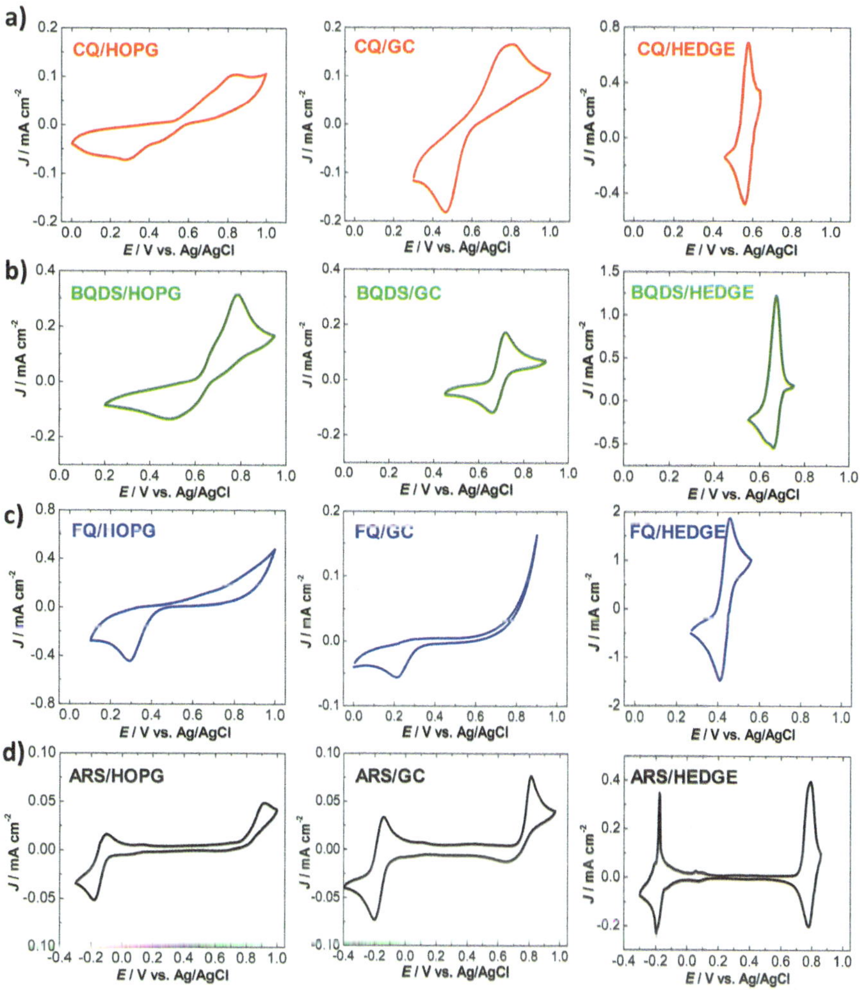

Fig. 4 Cyclic voltammograms for **a** CQ, **b** BQDS, **c** FQ, and **d** ARS at acid media using GC, HOPG, and HEDGE electrodes. Reprinted (adapted) with permission from Ref. [51]. Copyright 2021 American Chemical Society

2.2 *Flavin-Based Redox Flow Battery*

Another interesting class of organic molecules used in bioinspired batteries is flavins (Fig. 6), which act as a cofactor in several enzyme-catalyzed biological reactions [9, 52]. Flavins are heterocyclic compounds that can undergo a redox reaction by transferring one or two electrons and reacting with oxygen, which includes the formation of adducts [53, 54]. Riboflavin, which is composed of an isoalloxazine unit, is vitamin B_2 [52]. Furthermore, in addition to the basic structure of riboflavin, flavin

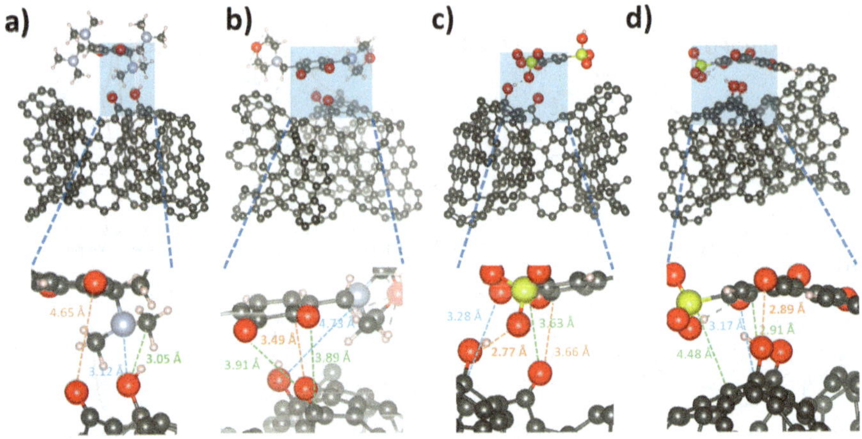

Fig. 5 Molecular dynamics simulations of the quinone-HEDGE interaction for **a** FQ, **b** CQ, **c** BQDS, and **d** ARS. The gray, yellow, red, and blue spheres denote carbon, hydrogen, sulfur, and oxygen atoms, respectively. Reprinted (adapted) with permission from Ref [51]. Copyright 2021 American Chemical Society

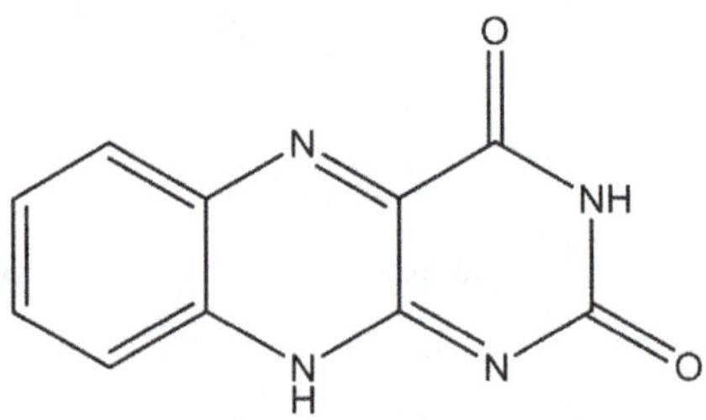

Fig. 6 Chemical structure of flavin

mononucleotide (FMN) and flavin adenine dinucleotide are also formed by the planar isoalloxazine ring [42]. This type of organic compound is versatile because of the various redox states that the flavin ring can adopt [52], and it can be dissolved in water, fat, blood, as well as in biological systems [42]; thus, flavins are promising materials for bioinspired-RFB development.

Orita et al. [42] developed a bioinspired aqueous RFB using flavin mononucleotide sodium salt (FMN-Na) as the negative electrolyte and $K_4[Fe(CN)_6]$ as the positive electrolyte. Although FMN-Na has the highest hydro solubility compared to other flavins, nicotinamide was used as a hydrotropic agent in its electrolyte to further increase its solubility. This high solubility is due to FMN-Na having a higher polarity in alkaline solutions than in acidic or neutral solutions. The polarity is caused by the large negative charge (FMN^{3-}), resulting in the most stable solvated state by water. The half-cell study showed that FMN-Na responds better in alkaline media, since the CV displayed higher current densities and lower peak separations.

The electrochemistry of the battery was tested at both low and high concentrations. First, the RFB was assembled using 0.06 mol L^{-1} FMN-Na and 0.1 mol L^{-1} $K_4[Fe(CN)_6]$ in 1 mol L^{-1} KOH, which gives a theoretical capacity of 1.34 Ah L^{-1}. The charge–discharge curves were obtained in galvanostatic mode using current densities of 5, 10, and 20 mA cm^{-2}. Figure 7a shows the obtained results. The initial discharge capacity at 10 mA cm^{-2} was 1.31 Ah L^{-1}, which is close to the theoretical value. Figure 7b shows the 1st, 100th, and 200th cycles of the galvanostatic charge–discharge measurement at 10 mA cm^{-2}. Over 200 cycles, stable discharge capacities and an increase in CE were observed (Fig. 7c). However, there are two

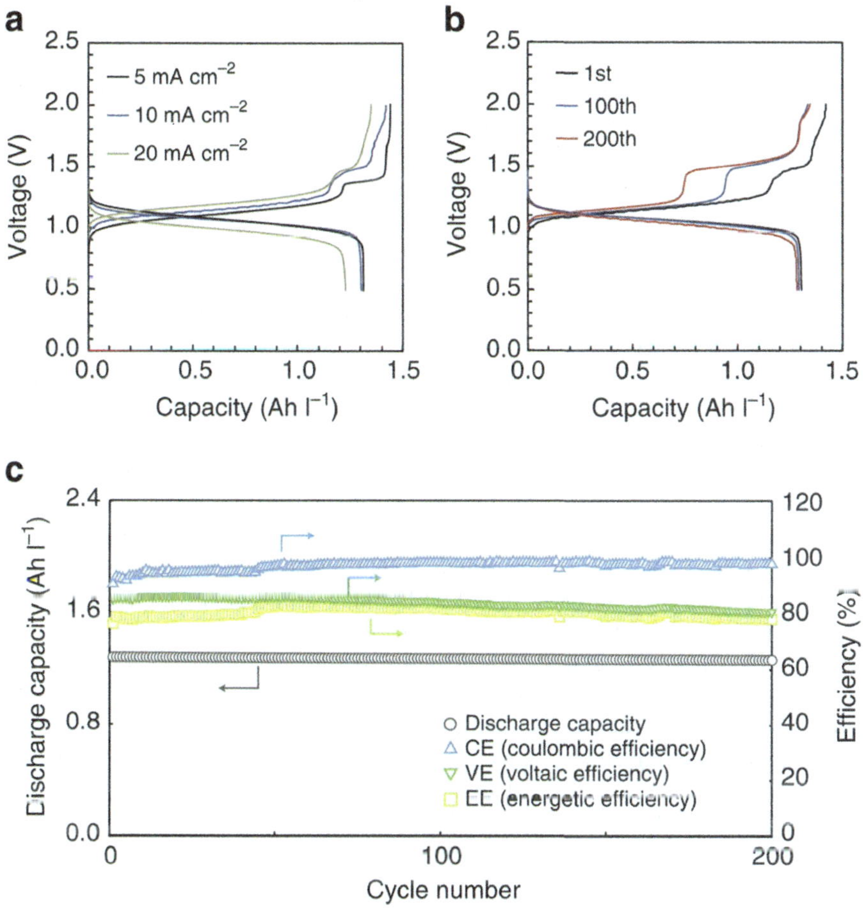

Fig. 7 Electrochemical performance of an RFB assembled with 0.1 mol L^{-1} $K_4[Fe(CN)_6]$ and 0.06 mol L^{-1} FMN-Na in 1 mol L^{-1} KOH. **a** Charge–discharge curves at 5, 10, and 20 mA cm^{-2}. **b** Charge–discharge at 1st, 100th, and 200th cycles at 10 mA cm^{-2}. **c** Discharge capacity and coulombic, voltaic, and energetic efficiencies for the assembled $K_4[Fe(CN)_6]$/FMN-Na RFB at 10 mA cm^{-2}. Reproduced with permission from Ref. [42] with the copyright of Springer Nature

plateaus, at 1.15 and 1.50 V, in the load curve, which suggests a reduction of the FMN-Na monomer and dimer.

Then, the concentration of the electrolytes was increased, and a new RFB was studied, using 0.4 mol L^{-1} $K_4[Fe(CN)_6]$ in 1 mol L^{-1} KOH and 0.24 mol L^{-1} FMN-Na and 1 mol L^{-1} nicotinamide in 1 mol L^{-1} KOH. The charge–discharge profile (Fig. 8a) was analyzed at current densities of 10–80 mA cm^{-2}. The two charge plateaus previously observed were only seen at 10 and 25 mA cm^{-2}. Thus, lower charge capacities were observed at higher current densities. The initial capacity obtained at 25 mA cm^{-2} was 5.03 Ah L^{-1}, which is close to the theoretical value for the RFB (5.36 Ah L^{-1}). Charge–discharge curves, cycling discharge capacities, and efficiencies were obtained over 200 cycles, lasting 76 h at current densities of 50 and 80 mA cm^{-2} (Fig. 8b, c). Figure 8c shows that 99% of the discharge capacity was retained after 100 cycles, and the CE was greater than 99% at 80 mA cm^{-2}. Figure 8d shows the power curve, indicating a peak power density equal to 0.16 W cm^{-2} at 0.3 A cm^{-2}, which is greater than that of the VRFB (0.12 W cm^{-2} at 0.15 A cm^{-2}).

Orita et al. [42] proposed that resonance structures in the reduced state allow for stable redox cycle performance. They also project new studies with other positive electrolytes that have even greater solubility in water. They concluded that FMN-Na is a promising active material for RFBs, in part because it is highly abundant, and not harmful to the environment.

Another study using flavin derivatives was published by Lin et al. [41]. It reported alloxazine-based, meaning a tautomer of the isoalloxazine structure of vitamin B_2, active material on the negative side and a mixture of $K_4[Fe(CN)_6]/K_3[Fe(CN)_6]$ on the positive side. However, because alloxazine has low solubility, the alloxazine core was functionalized with an alkaline-soluble carboxylic acid group by reacting o-phenylenediamine-4-carboxylic acid with alloxan to produce an isomeric mixture of alloxazine 7/8-carboxylic acid (ACA). Then, an aqueous RFB was assembled using 0.5 mol L^{-1} ACA as a negative electrolyte and 0.4 mol L^{-1} $K_4[Fe(CN)_6]$ + 40 mmol L^{-1} $K_3[Fe(CN)_6]$ as a positive electrolyte; both electrolytes were adjusted to pH 14 in KOH. This cell had an OCV of 1.2 V. Looking at the polarization curve, it is possible to observe a peak power density of 0.35 W cm^{-2} at a current density of 0.58 A cm^{-2}. Additionally, after 400 cycles, a current efficiency of 99.7% at 0.1 A cm^{-2} was achieved; however, the energy efficiency is only 63% (presumably due to an increase in system resistance), while the capacity-retention rate is greater than 91%. Therefore, considering new systems, they simulated, through theoretical modeling, two other molecules, also derived from alloxazine, which indicates an additional increase of almost 10% in the battery voltage. Thus, these bioinspired molecules offer a promising avenue for the development of AORFBs.

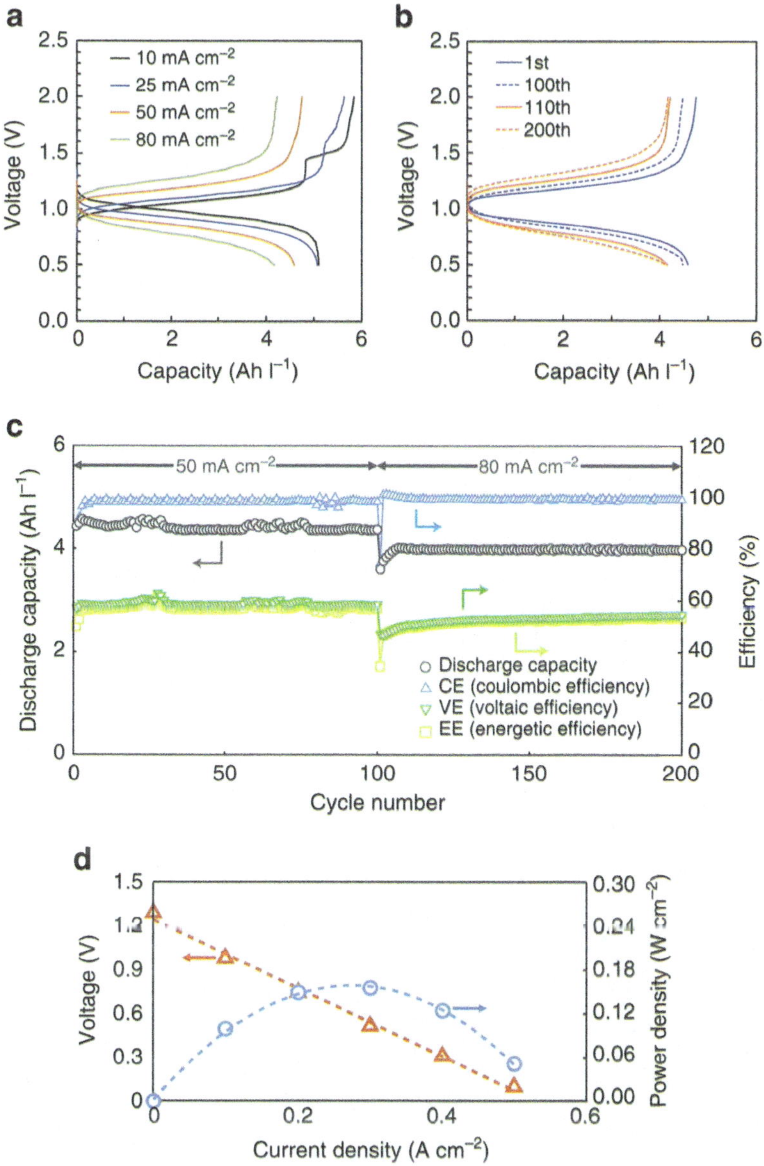

Fig. 8 Electrochemical performance of an RFB assembled with 0.4 mol L^{-1} $K_4[Fe(CN)_6]$ and 0.24 mol L^{-1} FMN-Na in 1 mol L^{-1} KOH and 1 mol L^{-1} NA. **a** Charge–discharge curves at 10, 25, 50, and 80 mA cm^{-2}. **b** Charge–discharge curves at 1st and 100th cycles at 50 mA cm^{-2}, and 110th and 200th cycles at 80 mA cm^{-2}. **c** Discharge capacity and coulombic, voltaic, and energetic efficiencies over 200 cycles. **d** Cell voltage vs current density vs power density for a high concentration of $K_4[Fe(CN)_6]$ / FMN-Na RFB. Reproduced with permission from Ref. [42] with the copyright of SpringerNature

3 Bioinspired Electrolytes: Using Biopolymers to Produce Greener and Safer Electrolytes

Wearable, ingestible, and implantable devices are currently at the cutting-edge of research and are applicable to many areas of technology, including healthcare [5, 55]. Although these technologies present many advantages, there is a lack of reliable and safe energy-storage components that meet the requirements of these devices [5]. Lithium-based batteries and silver-silver oxide are currently most often used as power sources in these devices, but they present concerns that include gastric juice pH enhancement [55, 56], toxic electrolyte leakage [57], and skin burning [58]. Thus, the development of reliable and safe batteries is an important step in the further development of these technologies.

Hydrogels are defined as 3D cross-linked networks capable of absorbing water without dissolving, and exhibit viscoelastic rheological behavior [59, 60]. Figure 9a depicts the basic structure of a hydrogel, where the blue chains represent the polymer chains, and the red dashed lines represent the cross-linking interactions between them. These materials possess unique characteristics, such as high water content and mechanical stability, and can achieve high conductivities, which makes them appropriate materials for the next generation of aqueous batteries [61]. Furthermore, the use of semi-solid electrolytes allows manufacturing power sources with different shapes, which is an essential requirement for emerging technologies, but is not feasible for current rigid batteries [5, 62].

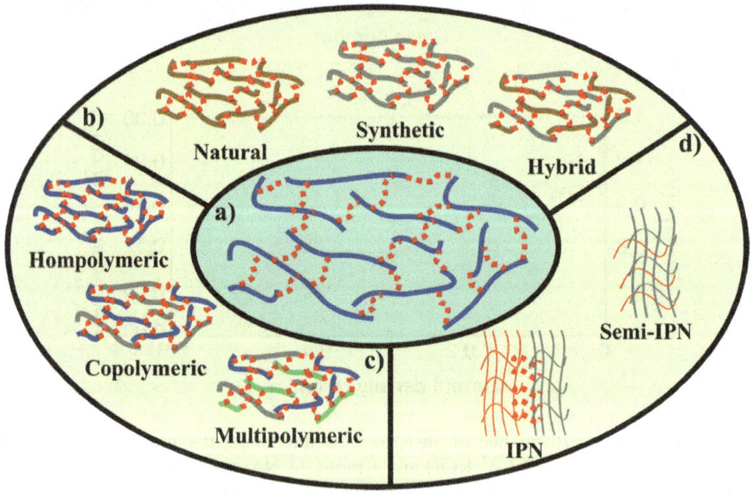

Fig. 9 Illustration of hydrogel microstructure and some of its classifications. **a** Basic microstructure of a hydrogel. **b** and **c** classification according to the origin of the polymer (or monomer) and the number of polymers used in the hydrogel synthesis, respectively. **d** Classification according to the interaction of different polymer matrixes

There are many ways to classify hydrogels according to their structure, polymer origin, and function [60, 61, 63, 64]. As to the origin of the polymers (or monomers) applied in the synthesis procedure, hydrogels can be natural-derived, synthetic, or hybrid, when a combination of natural and synthetic is used (Fig. 9b) [61, 63]. Examples of natural, synthetic, and hybrid hydrogels include gelatin [65], polyacrylic acid (PAA) [66], and carboxymethyl cellulose with polyethylene oxide networks [67]. Homopolymeric hydrogels are produced using only one polymer, copolymeric uses two different polymers, and multipolymeric uses three or more polymers, as shown in Fig. 9c [64]. Copolymeric hydrogels can be further classified into semi-interpenetrating (semi-IPN) and interpenetrating (IPN) networks. Semi-IPN occurs when some linear single chains of one polymer penetrate the porous structure of the cross-linked polymer. In contrast, IPN uses two polymeric cross-linked networks of different polymers that interact with each other by physical means [61, 63]. Figure 9d presents a sketch of the IPN and semi-IPN hydrogels. Both IPN and semi-IPN hydrogels have been employed as electrolytes in ESDs development [68, 69]. In terms of the charge content of the polymer backbone or pendant groups, hydrogels are classified as neutral, anionic, cationic, or ampholytic, when both cationic and anionic charges are present [61, 63, 64]. Finally, these materials can be classified according to their multifunctional properties, such as self-healing [70] and ionic conductivity [71].

The 3D hierarchical structure of hydrogels is built by physical or chemical means [72]. Physical hydrogels use physical interactions, such as electrostatic interactions [73], hydrogen bonding [73, 74], π–π stacking [75], and hydrophobic interactions [74] as the driving force to achieve the gel state. Chemical hydrogels are synthesized by covalently bonding the polymer chains, which can be achieved by free-radical polymerization [76], the addition of a cross-linker agent [77], condensation reactions between organic functions [78], and click chemistry [79, 80]. Since covalent bonds are stronger than physical interactions, these hydrogels have improved mechanical properties [64]. However, the use of some cross-linkers [81, 82] enhances hydrogel toxicity, which is a concern in applications where biocompatibility is mandatory.

The next step in energy-storage technology is to develop new technologies thinking in the entire production chain, that is, from device manufacturing through to disposal or recycling [83]. Natural polymers are extracted from natural sources (e.g., alginate, chitosan, and agarose) and have key characteristics such as renewability, biocompatibility, biodegradability, and low cost [10]. Moreover, their aqueous solubility and wide range of organic functions make these polymers suitable for the construction of natural hydrogels which are also biocompatible and biodegradable [10, 59]. These characteristics make natural hydrogels appropriate materials for the next generation of ESDs, with a view to meeting the requirements of incipient technologies and addressing current environmental issues. The next section reviews recent advances in the application of natural-based hydrogels for the development of reliable and safe batteries.

3.1 Batteries Employing Natural-Based Hydrogels as Electrolyte

Researchers around the world have been searching for alternative materials to manufacture ESDs that meet the safety and flexibility requirements of wearable, ingestible, and implantable devices [4, 5, 84]. Since hydrogels possess high water content [85], and can achieve high ionic conductivities [86, 87], they can replace liquid aqueous electrolytes, which make the battery safer [88] and allow it to be produced in different shapes [65, 70], including flexible ones [11].

Crespilho et al. [89] obtained insights into the RFB research field and used previously reported molecules [36] incorporated in an easily synthesized agarose-based hydrogel to produce an organic/organometallic bioinspired secondary microbattery (MB). To produce the hydrogel electrolytes, bis(3-trimethylammonio) propyl viologen (BTMAP-Vi) and bis((3-thimethylammonio)propyl) ferrocene (BTMAP-Fc) were incorporated in 1.5% (w/w) agarose in 1 mol L^{-1} KCl solution and heated to 90 °C, after which the hot solutions were dripped onto flexible carbon fiber electrode surfaces. Gelation then occurred at room temperature [89].

The MB was assembled by placing a piece of previously hydrated Selemion® DSV membrane on the hydrogel/electrode surface, the union of cathodic and anodic chambers, and sealing with silicone resin. Figure 10a shows the assembled MB. To evaluate the maximum capacity of the MB, it was potentiostically charged at 1.10 V until reaching an SOC of 92%, and then discharged at 0.30 V. The total discharged capacity was calculated as 0.79 mAh (Fig. 10b). At an SOC of 20%, the MB could supply 20 $\mu A\ cm^{-2}$ for 100 h, which demonstrates a volumetric capacity of 0.021 $mAh\ cm^{-2}\ \mu m^{-1}$. Figure 10c shows the results of the galvanostatic charge–discharge cycling at 40 mA cm^{-2}. The charging and discharging capacities showed almost no change over 100 cycles, giving a CE ranging from 94 to 97%. In addition to its electrochemical characteristics, the MB exhibited attractive toxicological and structural characteristics. It is non-flammable, since it uses water as the solvent, and possesses low toxicity, because the quantities of BTMAP-Vi and BTMAP-Fc employed are low to cause harmful effects [89]. Thus, the electrochemical, structural, and toxicological characteristics make this battery suitable for low-power consumption medical devices, such as ingestible pills [90].

Chitosan (CS) and polyvinyl alcohol (PVA), both biodegradable polymers, were employed in the synthesis of an alkaline electrolyte for application in a zinc-electrolytic manganese dioxide (Zn-EMD) battery. First, CS and PVA solutions were mixed in a 1:0.2 wt% ratio (CP solution) and poured into a silicone mold. Next, the CP solution was dried at 37 °C for 1200 min. Finally, the dried hydrogels were soaked in KOH solutions at different concentrations and at different times. The best properties were achieved by the hydrogel soaked in 5 M KOH solution for 45 min, known as {5}CP (Fig. 10d), which achieved the highest swelling ratio, highest conductivity (457.19 mS cm^{-1}), and a lower degree of crystallinity. This hydrogel was then used as a gel polymer electrolyte in a Zn-EMD battery [91].

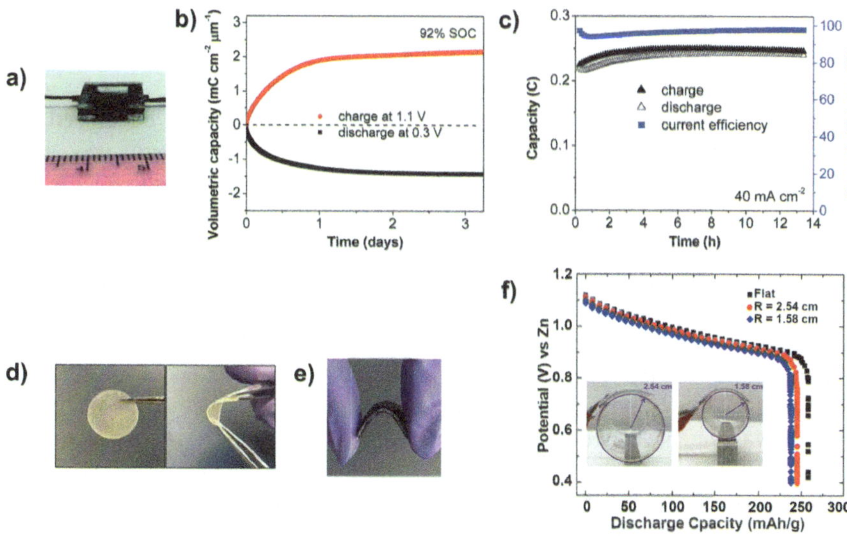

Fig. 10 Semi-solid batteries produced with polysaccharides. **a** Photograph, **b** potentiostatic charge discharge curves at 1.10 (red) and 0.30 **V** (black), and **c** Charge and discharge capacities and CE over 100 cycles at 40 mA cm^{-2} for microbattery built up with BTMAP-Fc and BTMAP-Vi. **d** Photograph of CP samples, **e** photograph of Zn-EMD flexible battery, and **f** galvanostatic discharge for the Zn-EMD battery under two bent conditions. Reproduced from Ref. [89] with permission from Royal Society of Chemistry, Reprinted (adapted) with permission from Ref. [91]. Copyright 2021 American Chemical Society

Figure 10e shows the produced Zn-EMD battery. Pre-treated flexible carbon cloths (CC) were used as current collectors in both cathode and anode. The battery was assembled using a metallic zinc foil as the anode, CC coated with EMD ink as the cathode (load of ~1.4 mg cm^{-2}), and {5}CP as both electrolyte and separator. In order to achieve better cathode-electrolyte contact, {5}CP was previously dipped in CP solution until becoming gooey and then placed on the cathode surface [91]. The Zn-EMD battery was further characterized by galvanostatic charge–discharge at 0.1 A g^{-1} at two different potential windows, that is, 0.4–1.2 V (low potential) and 0.4–1.6 V (high potential), chosen according to the cathode degradation potential. The low and high potential windows presented a specific capacity of 310 and 287 mAh g^{-1} after the first cycle and capacity retention of 74% and 17.4% after 50 cycles, respectively. These results are attributed to EMD's irreversible phase transition to δ-MnO_2, which leads to an enhancement of the battery impedance and prevents charge transfer between the electrode and electrolyte. Furthermore, energy dispersive x-ray analysis shows a higher content of $ZnMn_xO_y$ complexes after further cycling at the higher voltage window, which also contributes to the inferior performance of the battery cycled at this potential window. The Zn-EMD battery was further galvanostatic cycled at 0.5 A g^{-1} and showed an average specific capacity of 175 mAh g^{-1} over 300 cycles with a CE of 96.5%. However, at this current density, the battery presented a steady decrease in capacity during the first 15 cycles. Finally, to demonstrate the battery's flexibility, it was electrochemically characterized in two bent conformations and there was almost no change in the electrochemical performance, as shown in Fig. 10f [91].

Similarly, an alginate-based glycerol-plasticized hydrogel was developed and applied to zinc-iodide (Zn-I_2) batteries [71]. First, 0.75 g of glycerol was added to 50 mL of 30 mg mL^{-1} sodium alginate solution, with further stirring for 1 h. The resulting solution was cast on the cathode surface, degassed, and dried at 60 °C for 12 h. Finally, it was soaked in 1 mol L^{-1} zinc sulfate or acetate solution for 4 h. Figure 11a shows photographs of the alginate-based hydrogel. It could be bent or twisted without damage and even bear a weight of 200 g, which demonstrates its mechanical robustness and flexibility. The linear sweep voltammograms and electrochemical impedance spectroscopy measurements showed an operational voltage window of ~2.60 V and conductivity of 32.8 mS cm^{-1}, respectively. Finally, the hydrogel was capable of preventing zinc corrosion and I_2 shuttling, two mechanisms of Zn-I_2 battery failure; this affords a longer life for the novel batteries [71].

The semi-solid Zn-I_2 battery was assembled in a CR2025 shape with a zinc foil as the anode, the alginate-based hydrogel as the electrolyte, and a graphite paper coated with I_2@AC composite as the cathode. Galvanostatic charge–discharge at 0.2 A g^{-1} measurements with the $Zn(Ac)_2$-soaked alginate hydrogel showed a specific capacity, capacity retention, and CE of 183.4 mAh g^{-1} (Fig. 11b), 97.6%, and 99.3% over 200 cycles, respectively. For comparison, the authors also reported a Zn-I_2 battery with fiberglass soaked in $ZnSO_4$ as the separator. This battery exhibited a significantly inferior electrochemical performance (blue line in Fig. 9b). The galvanostatic charge–discharge assay at 1 A g^{-1} (Fig. 11c) shows capacity retention and CE of 66.8 and 99.5%, respectively, which demonstrates the long life of the battery and the shielding effects of the hydrogel over the zinc anode. Finally, a flexible pouch battery was assembled using this approach. This battery shows an OCV of 1.25 V, capacity of 3.0–3.5 mAh, and worked properly even in bent conformations [71].

Self-healing hydrogels can repair themselves after damage and have been suggested as an alternative to enhance the lifespan of batteries. In this context, a copolymeric hydrogel based on alginate and PAA was synthesized and applied as an electrolyte to build an aqueous lithium-ion yarn battery (ALIYB). To synthesize the hydrogel, PANa-Ca-SA, a solution of acrylic acid and bistrimethanesulfonimide lithium salt was neutralized with NaOH in an ice bath. Then, alginate, calcium chloride, and ammonium persulfate were dissolved in the solution and the resulting solution was kept at 40 °C for 30 h to allow polymerization. Calcium ions play a key role in this hydrogel as they electrostatically interact with PAA and alginate polymers and act as cross-linkers of the two polymeric networks. In addition, the electrostatic interaction between calcium ions and the polymers explains the remarkable self-healing property of the hydrogel, because it can easily be broken and healed (Fig. 11d). The authors investigated the influence of the cutting/self-healing cycles on the electrical and mechanical properties of the hydrogel. The results showed a small decrease in the PANa-Ca-SA conductivity, which was 88% of its initial value after the eighth cutting/self-healing cycle. In contrast, the mechanical properties showed a pronounced change. The original hydrogel could be stretched at 350% strain and

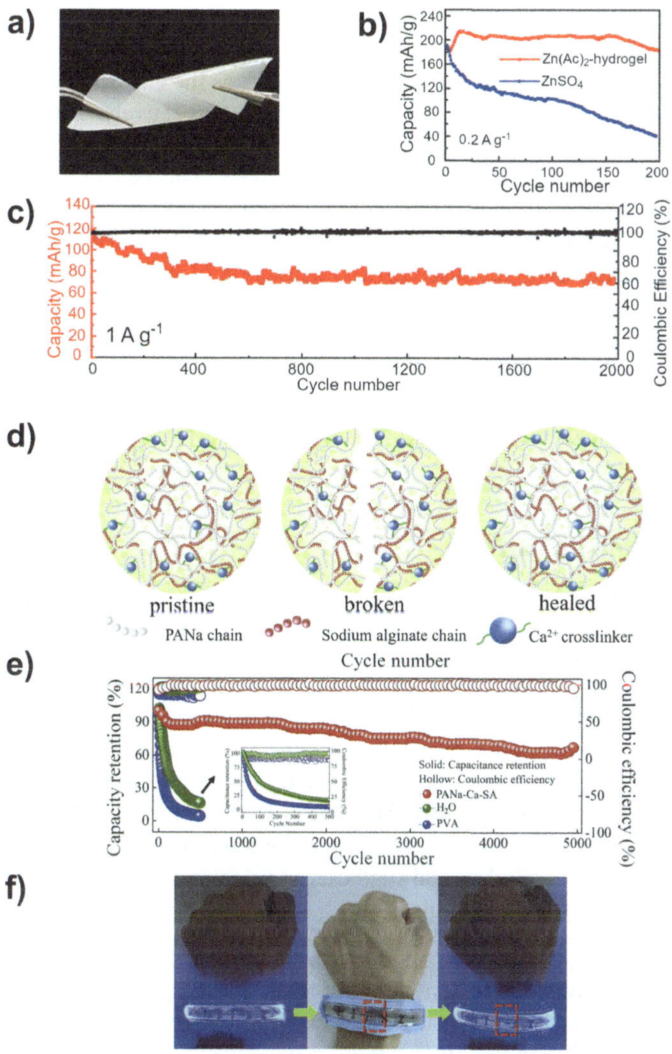

Fig. 11 Batteries produced using hybrid alginate-based hydrogel as electrolyte. **a** Photograph of twisted alginate-based hydrogel electrolyte, **b** discharge capacity at 0.2 A g^{-1} for Zn/I_2 batteries using the alginate-based hydrogel (red line) and aqueous $Zn(Ac)_2$ solution as electrolytes, respectively. **c** Capacity (red line) and CE (black line) for Zn/I_2 battery assembled with alginate-based electrolyte under galvanostatic cycling at 1 A g^{-1} over 2000 cycles. **d** Representation of pristine, broken and healed PANa-Ca-SA hydrogel. **e** Electrochemical performances of ALIYBs produced using PANa-Ca-SA hydrogel (red), water (green) and, PVA-based hydrogel (blue) electrolytes over 5000 cycles. **f** Photograph of a light strip powered by four ALIYBs before cutting (left), after cut (middle), and after self-healing (right). Reprinted (adapted) with permission from Ref. [71]. Copyright 2021 American Chemical Society. Reprinted from Energy-storage materials, v.28, Ji, Z et al., A both microscopically and macroscopically intrinsic self-healing long lifespan yarn battery, p. 334–341, Copyright (2021), with permission from Elsevier

235 kPa tensile strength, while for the eight-cut hydrogel these values were reduced to 48% and 45 kPa [70].

The ALIYB was assembled using PANa-Ca-SA as the electrolyte and $LiMn_2O_4$ and LiV_3O_8 coated yarns as the anode and cathode, respectively. The galvanostatic charge–discharge data at 0.5 A g^{-1} (Fig. 11e) show capacity retention and CE of 70% and nearly 100% after 5000 cycles, respectively. This behavior is much better than that of water and PVA-hydrogel-based lithium-ion batteries and is attributed to the self-healing capacity of the hydrogel, which prevents damage to the surfaces of the electrodes caused by the insertion and extraction of lithium ions. Furthermore, the ALIYB showed a healing efficiency of 68% and almost no change in its electrochemical performance after eight cutting/self-healing cycles. Figure 11f shows the application of four parallel coupled batteries to power a light strip and evidence the self-healing behavior of the battery, because the device turns off when the battery breaks and shines again after the battery has healed [70].

4 Final Considerations

Electrochemical ESDs are indispensable for the ongoing sustainable development of humankind because they are highly efficient and address the current environmental concerns. Moreover, nature-developed pathways and materials have inspired researchers to confront the global problems that arise from the high consumption of non-renewable energy sources. In the energy-research field, organic-based RFBs that use quinones and flavins; and flexible batteries that use natural-based gel electrolytes, are key examples of this integration. However, both technologies are in their infancy and their advancement will require a multidisciplinary effort in areas such as materials science, chemical synthesis and electrosynthesis, electrochemistry, polymer science, soft materials, quantum chemistry, and engineering, among others.

Acknowledgements T. B. and L. C. I. F. acknowledge São Paulo Research Foundation (FAPESP) for the fellowships 2020/03681-2 and 2019/21089-6. F. N. C. acknowledges FAPESP for the grants 2018/22214-6, 2019/15333-1, and 2019/12053-8. J. E. S. C. acknowledges Coordenação de Aperfeiçoamento de Pessoal de Nível Superior (CAPES)—Finance Code 1806336.

References

1. Yang, Z., Zhang, J., Kintner-Meyer, M.C.W., et al.: Electrochemical energy storage for green grid. Chem. Rev. **111**, 3577–3613 (2011). https://doi.org/10.1021/cr100290v
2. Nguyen, T., Savinell, R.F.: Flow batteries. Electrochem. Soc. Interface **19**, 54–56 (2010). https://doi.org/10.1149/2.F06103if
3. Koydemir, H.C., Ozcan, A.: Wearable and implantable sensors for biomedical applications. Annu. Rev. Anal. Chem. **11**, 127–146 (2018). https://doi.org/10.1146/annurev-anchem-061417-125956

4. Ling, Y., An, T., Yap, L.W., et al.: Disruptive, soft, wearable sensors. Adv. Mater. **32**, 1904664 (2020). https://doi.org/10.1002/adma.201904664
5. Sunwoo, S.-H., Ha, K.-H., Lee, S., et al.: Wearable and implantable soft bioelectronics: device designs and material strategies. Annu. Rev. Chem. Biomol. Eng. **12**, 359–391 (2021). https://doi.org/10.1146/annurev-chembioeng-101420-024336
6. Whitesides, G.M., Whitesides, G.M.: Bioinspiration: something for everyone (2015)
7. Monks, T.J., Hanzlik, R.P., Cohen, G.M., et al.: Quinone chemistry and toxicity. Toxicol. Appl. Pharmacol. **112**, 2–16 (1992). https://doi.org/10.1016/0041-008X(92)90273-U
8. Son, E.J., Kim, J.H., Kim, K., Park, C.B.: Quinone and its derivatives for energy harvesting and storage materials. J. Mater. Chem. A **4**, 11179–11202 (2016). https://doi.org/10.1039/C6TA03123D
9. Massey, V.: The chemical and biological versatility of riboflavin. Biochem. Soc. Trans. **28**, 283–296 (2000). https://doi.org/10.1042/bst0280283
10. Cui, C., Fu, Q., Meng, L., et al.: Recent progress in natural biopolymers conductive hydrogels for flexible wearable sensors and energy devices: materials, structures, and performance. ACS Appl. Bio. Mater. **4**, 85–121 (2021). https://doi.org/10.1021/acsabm.0c00807
11. Cheng, X., Pan, J., Zhao, Y., et al.: Gel polymer electrolytes for electrochemical energy storage. Adv. Energy Mater. **8**, 1–16 (2018). https://doi.org/10.1002/aenm.201702184
12. Prifti, H., Parasuraman, A., Winardi, S., et al.: Membranes for redox flow battery applications. Membranes (Basel) **2**, 275–306 (2012). https://doi.org/10.3390/membranes2020275
13. Winsberg, J., Hagemann, T., Janoschka, T., et al.: Redox-flow batteries: from metals to organic redox-active materials. Angew. Chemie Int. Ed. **56**, 686–711 (2017). https://doi.org/10.1002/anie.201604925
14. Leung, P., Shah, A.A., Sanz, L., et al.: Recent developments in organic redox flow batteries: a critical review. J. Power Sour. **360**, 243–283 (2017). https://doi.org/10.1016/j.jpowsour.2017.05.057
15. Sánchez-Díez, E., Ventosa, E., Guarnieri, M., et al.: Redox flow batteries: status and perspective towards sustainable stationary energy storage. J. Power Sour. **481**, 228804 (2021). https://doi.org/10.1016/j.jpowsour.2020.228804
16. Alotto, P., Guarnieri, M., Moro, F.: Redox flow batteries for the storage of renewable energy: a review. Renew. Sustain. Energy Rev. **29**, 325–335 (2014). https://doi.org/10.1016/j.rser.2013.08.001
17. Kazacos, M., Cheng, M., Skyllas-Kazacos, M.: Vanadium redox cell electrolyte optimization studies. J. Appl. Electrochem. **20**, 463–467 (1990). https://doi.org/10.1007/BF01076057
18. Rahman, F., Skyllas-Kazacos, M.: Vanadium redox battery: positive half-cell electrolyte studies. J. Power Sour. **189**, 1212–1219 (2009). https://doi.org/10.1016/j.jpowsour.2008.12.113
19. Rahman, F., Skyllas-Kazacos, M.: Solubility of vanadyl sulfate in concentrated sulfuric acid solutions. J. Power Sour. **72**, 105–110 (1998). https://doi.org/10.1016/S0378-7753(97)02692-X
20. Huskinson, B., Marshak, M.P., Suh, C., et al.: A metal-free organic–inorganic aqueous flow battery. Nature **505**, 195–198 (2014). https://doi.org/10.1038/nature12909
21. Gerhardt, M.R., Tong, L., Gómez-Bombarelli, R., et al.: Anthraquinone derivatives in aqueous flow batteries. Adv. Energy Mater. **7**, 1601488 (2017). https://doi.org/10.1002/aenm.201601488
22. Yang, Z., Tong, L., Tabor, D.P., et al.: Alkaline benzoquinone aqueous flow battery for large-scale storage of electrical energy. Adv. Energy Mater. **8**, 1702056 (2018). https://doi.org/10.1002/aenm.201702056
23. Lin, K., Chen, Q., Gerhardt, M.R., et al.: Alkaline quinone flow battery. Science **349**(80), 1529–1532 (2015). https://doi.org/10.1126/science.aab3033
24. Er, S., Suh, C., Marshak, M.P., Aspuru-Guzik, A.: Computational design of molecules for an all-quinone redox flow battery. Chem. Sci. **6**, 885–893 (2015). https://doi.org/10.1039/C4SC03030C
25. Bachman, J.E., Curtiss, L.A., Assary, R.S.: Investigation of the redox chemistry of anthraquinone derivatives using density functional theory. J. Phys. Chem. A **118**, 8852–8860 (2014). https://doi.org/10.1021/jp5060777

26. Lee, W., Park, G., Kwon, Y.: Alkaline aqueous organic redox flow batteries of high energy and power densities using mixed naphthoquinone derivatives. Chem. Eng. J. **386**, 123985 (2020). https://doi.org/10.1016/j.cej.2019.123985
27. Yang, B., Hoober-Burkhardt, L., Wang, F., et al.: An inexpensive aqueous flow battery for large-scale electrical energy storage based on water-soluble organic redox couples. J. Electrochem. Soc. **161**, A1371–A1380 (2014). https://doi.org/10.1149/2.1001409jes
28. Lima, A. R.F., Pereira, R.C., Azevedo, J., et al.: On the path to aqueous organic redox flow batteries: Alizarin red S alkaline negolyte. Performance evaluation and photochemical studies. J. Mol. Liq. **336**, 116364 (2021). https://doi.org/10.1016/j.molliq.2021.116364
29. Zhou, M., Chen, Y., Salla, M., et al.: Single-molecule redox-targeting reactions for a pH-neutral aqueous organic redox flow battery. Angew. Chem. Int. Ed. **59**, 14286–14291 (2020). https://doi.org/10.1002/anie.202004603
30. Lee, W., Permatasari, A., Kwon, B.W., Kwon, Y.: Performance evaluation of aqueous organic redox flow battery using anthraquinone-2,7-disulfonic acid disodium salt and potassium iodide redox couple. Chem. Eng. J. **358**, 1438–1445 (2019). https://doi.org/10.1016/j.cej.2018.10.159
31. Guiheneuf, S., Lê, A., Godet-Bar, T., et al.: Behaviour of 3,4-dihydroxy-9,10-anthraquinone-2-sulfonic acid in alkaline medium: towards a long-cycling aqueous organic redox flow battery. ChemElectroChem **8**, 2526–2533 (2021). https://doi.org/10.1002/celc.202100284
32. Hu, B., Luo, J., Hu, M., et al.: A pH-neutral, metal-free aqueous organic redox flow battery employing an ammonium anthraquinone anolyte. Angew. Chem. **131**, 16782–16789 (2019). https://doi.org/10.1002/ange.201907934
33. Lai, Y.Y., Li, X., Liu, K., et al.: Stable low-cost organic dye anolyte for aqueous organic redox flow battery. ACS Appl. Energy Mater. **3**, 2290–2295 (2020). https://doi.org/10.1021/acsaem.9b01735
34. Liu, T., Wei, X., Nie, Z., et al.: A total organic aqueous redox flow battery employing a low cost and sustainable methyl viologen anolyte and 4-HO-TEMPO catholyte. Adv. Energy Mater. **6**, 1501449 (2016). https://doi.org/10.1002/aenm.201501449
35. Liu, L., Yao, Y., Wang, Z., Lu, Y.-C.: Viologen radical stabilization by molecular spectators for aqueous organic redox flow batteries. Nano Energy **84**, 105897 (2021). https://doi.org/10.1016/j.nanoen.2021.105897
36. Beh, E.S., De Porcellinis, D., Gracia, R.L., et al.: A neutral pH aqueous organic-organometallic redox flow battery with extremely high capacity retention. ACS Energy Lett. **2**, 639–644 (2017). https://doi.org/10.1021/acsenergylett.7b00019
37. Hu, B., DeBruler, C., Rhodes, Z., Liu, T.L.: Long-cycling aqueous organic redox flow battery (AORFB) toward sustainable and safe energy storage. J. Am. Chem. Soc. **139**, 1207–1214 (2017). https://doi.org/10.1021/jacs.6b10984
38. Tsehaye, M.T., Yang, X., Janoschka, T., et al.: Study of anion exchange membrane properties incorporating N-spirocyclic quaternary ammonium cations and aqueous organic redox flow battery performance. Membranes **11** (2021)
39. Xing, X., Huo, Y., Wang, X., et al.: A benzophenone-based anolyte for high energy density all-organic redox flow battery. Int. J. Hydrog. Energy **42**, 17488–17494 (2017)
40. Sreenath, S., Nayanthara, P.S., Pawar, C.M., et al.: Phenolic triamine dangling poly(VDF-co-HFP) anion exchange membrane for all aqueous organic redox flow battery. J. Energy Storage **40**, 102689 (2021). https://doi.org/10.1016/j.est.2021.102689
41. Lin, K., Gómez-Bombarelli, R., Beh, E.S., et al.: A redox-flow battery with an alloxazine-based organic electrolyte. Nat. Energy **1**, 1–8 (2016). https://doi.org/10.1038/nenergy.2016.102
42. Orita, A., Verde, M.G., Sakai, M., Meng, Y.S.: A biomimetic redox flow battery based on flavin mononucleotide. Nat. Commun. **7**, 1–8 (2016). https://doi.org/10.1038/ncomms13230
43. Quan, M., Sanchez, D., Wasylkiw, M.F., Smith, D.K.: Voltammetry of quinones in unbuffered aqueous solution: reassessing the roles of proton transfer and hydrogen bonding in the aqueous electrochemistry of quinones. J. Am. Chem. Soc. **129**, 12847–12856 (2007). https://doi.org/10.1021/ja0743083
44. Kim, K.C., Liu, T., Lee, S.W., Jang, S.S.: First-principles density functional theory modeling of li binding: thermodynamics and redox properties of quinone derivatives for lithium-ion batteries. J. Am. Chem. Soc. **138**, 2374–2382 (2016). https://doi.org/10.1021/jacs.5b13279

45. Xu, Y., Wen, Y., Cheng, J., et al.: Study on a single flow acid Cd-chloranil battery. Electrochem. Commun. **11**, 1422–1424 (2009). https://doi.org/10.1016/j.elecom.2009.05.021
46. Xu, Y., Wen, Y.-H., Cheng, J., et al.: A study of tiron in aqueous solutions for redox flow battery application. Electrochim. Acta **55**, 715–720 (2010). https://doi.org/10.1016/j.electacta.2009.09.031
47. Hofmann, J.D., Schröder, D.: Which parameter is governing for aqueous redox flow batteries with organic active material? Chem. Ing. Tech. **91**, 786–794 (2019). https://doi.org/10.1002/cite.201800162
48. Yang, B., Hoober-Burkhardt, L., Krishnamoorthy, S., et al.: High-performance aqueous organic flow battery with quinone-based redox couples at both electrodes. J. Electrochem. Soc. **163**, A1442–A1449 (2016). https://doi.org/10.1149/2.1371607jes
49. Wu, M., Jing, Y., Wong, A.A., et al.: Extremely stable anthraquinone negolytes synthesized from common precursors. Chemistry **6**, 1432–1442 (2020). https://doi.org/10.1016/j.chempr.2020.03.021
50. Jing, Y., Wu, M., Wong, A.A., et al.: In situ electrosynthesis of anthraquinone electrolytes in aqueous flow batteries. Green Chem. **22**, 6084–6092 (2020). https://doi.org/10.1039/D0GC02236E
51. Sedenho, G.C., De Porcellinis, D., Jing, Y., et al.: Effect of molecular structure of quinones and carbon electrode surfaces on the interfacial electron transfer process. ACS Appl. Energy Mater. **3**, 1933–1943 (2020). https://doi.org/10.1021/acsaem.9b02357
52. Miura, R.: Versatility and specificity in flavoenzymes: control mechanisms of flavin reactivity. Chem. Rec. **1**, 183–194 (2001). https://doi.org/10.1002/tcr.1007
53. Entsch, B., Ballou, D.P.: Flavins. In: Encyclopedia of Biological Chemistry, pp. 309–313. Elsevier (2013)
54. Fagan, R.L., Palfey, B.A.: Flavin-dependent enzymes. In: Comprehensive Natural Products II, pp. 37–113. Elsevier (2010)
55. Kalantar-Zadeh, K., Ha, N., Ou, J.Z., Berean, K.J.: Ingestible sensors. ACS Sens. **2**, 468–483 (2017). https://doi.org/10.1021/acssensors.7b00045
56. Samad, L., Ali, M., Ramzi, H.: Button battery ingestion: Hazards of esophageal impaction. J. Pediatr. Surg. **34**, 1527–1531 (1999). https://doi.org/10.1016/S0022-3468(99)90119-7
57. Chevin, J.C., Attik, G., Dika, H., et al.: Button battery induced cell damage: a pathophysiological study. Electrochem. Commun. **10**, 1756–1760 (2008). https://doi.org/10.1016/j.elecom.2008.09.002
58. Mankowski, P.J., Kanevsky, J., Bakirtzian, P., Cugno, S.: Cellular phone collateral damage: a review of burns associated with lithium battery powered mobile devices. Burns **42**, e61–e64 (2016). https://doi.org/10.1016/j.burns.2015.10.012
59. Catoira, M.C., Fusaro, L., Di Francesco, D., et al.: Overview of natural hydrogels for regenerative medicine applications. J. Mater. Sci. Mater. Med. **30**, 115 (2019). https://doi.org/10.1007/s10856-019-6318-7
60. Guo, Y., Bae, J., Zhao, F., Yu, G.: Functional hydrogels for next-generation batteries and supercapacitors. Trends Chem. **1**, 335–348 (2019). https://doi.org/10.1016/j.trechm.2019.03.005
61. Guo, Y., Bae, J., Fang, Z., et al.: Hydrogels and hydrogel-derived materials for energy and water sustainability. Chem. Rev. **120**, 7642–7707 (2020). https://doi.org/10.1021/acs.chemrev.0c00345
62. Yang, Z., Deng, J., Chen, X., et al.: A highly stretchable, fiber-shaped supercapacitor. Angew. Chem. Int. Ed. **52**, 13453–13457 (2013). https://doi.org/10.1002/anie.201307619
63. Thakur, V.K., Thakur, M.K.: Hydrogels: Recent Advances. Springer, Sinpagore (2018)
64. Ali, A., Ahmed, S.: Recent advances in edible polymer based hydrogels as a sustainable alternative to conventional polymers. J. Agric. Food Chem. **66**, 6940–6967 (2018). https://doi.org/10.1021/acs.jafc.8b01052
65. Park, J., Park, M., Nam, G., et al.: All-solid-state cable-type flexible zinc-air battery. Adv. Mater. **27**, 1396–1401 (2015). https://doi.org/10.1002/adma.201404639

66. Tran, T.N.T., Chung, H.-J., Ivey, D.G.: A study of alkaline gel polymer electrolytes for rechargeable zinc—Air batteries. Electrochim Acta **327**, 135021 (2019). https://doi.org/10.1016/j.electacta.2019.135021
67. Colò, F., Bella, F., Nair, J.R., et al.: Cellulose-based novel hybrid polymer electrolytes for green and efficient Na-ion batteries. Electrochim Acta **174**, 185–190 (2015). https://doi.org/10.1016/j.electacta.2015.05.178
68. Wang, P.-H., Tseng, L.-H., Li, W.-C., et al.: Zwitterionic semi-IPN electrolyte with high ionic conductivity and high modulus achieving flexible 2.4 V aqueous supercapacitors. J. Taiwan Inst. Chem. Eng. **126**, 58–66 (2021). https://doi.org/10.1016/j.jtice.2021.06.044
69. Yan, T., Zou, Y., Zhang, X., et al.: Hydrogen bond interpenetrated agarose/PVA network: a highly ionic conductive and flame-retardant gel polymer electrolyte. ACS Appl. Mater. Interfaces **13**, 9856–9864 (2021). https://doi.org/10.1021/acsami.0c20702
70. Ji, Z., Wang, H., Chen, Z., et al.: A both microscopically and macroscopically intrinsic self-healing long lifespan yarn battery. Energy Storage Mater. **28**, 334–341 (2020). https://doi.org/10.1016/j.ensm.2020.03.020
71. Shang, W., Zhu, J., Liu, Y., et al.: Establishing high-performance quasi-solid Zn/I_2 batteries with alginate-based hydrogel electrolytes. ACS Appl. Mater. Interfaces **13**, 24756–24764 (2021). https://doi.org/10.1021/acsami.1c03804
72. Varaprasad, K., Raghavendra, G.M., Jayaramudu, T., et al.: A mini review on hydrogels classification and recent developments in miscellaneous applications. Mater. Sci. Eng. C **79**, 958–971 (2017). https://doi.org/10.1016/j.msec.2017.05.096
73. Lin, Y., Zhang, H., Liao, H., et al.: A physically crosslinked, self-healing hydrogel electrolyte for nano-wire PANI flexible supercapacitors. Chem. Eng. J. **367**, 139–148 (2019). https://doi.org/10.1016/j.cej.2019.02.064
74. Xu, R., Ma, S., Lin, P., et al.: High strength astringent hydrogels using protein as the building block for physically cross-linked multi-network. ACS Appl. Mater. Interfaces **10**, 7593–7601 (2018). https://doi.org/10.1021/acsami.7b04290
75. Wang, H., Zhou, C., Zhu, H., et al.: Hierarchical porous carbons from carboxylated coal-tar pitch functional poly(acrylic acid) hydrogel networks for supercapacitor electrodes. RSC Adv. **10**, 1095–1103 (2020). https://doi.org/10.1039/C9RA09141F
76. Wang, Z., Mo, F., Ma, L., et al.: Highly compressible cross-linked polyacrylamide hydrogel-enabled compressible $Zn–MnO_2$ battery and a flexible battery-sensor system. ACS Appl. Mater. Interfaces **10**, 44527–44534 (2018). https://doi.org/10.1021/acsami.8b17607
77. Zerbinati, N., Esposito, C., Cipolla, G., et al.: Chemical and mechanical characterization of hyaluronic acid hydrogel cross-linked with polyethylen glycol and its use in dermatology. Dermatol. Ther. **33**, e13747 (2020). https://doi.org/10.1111/dth.13747
78. Koo, B., Kim, H., Cho, Y., et al.: A highly cross-linked polymeric binder for high-performance silicon negative electrodes in lithium ion batteries. Angew. Chem. Int. Ed. **51**, 8762–8767 (2012). https://doi.org/10.1002/anie.201201568
79. Li, Y., Wang, X., Han, Y., et al.: Click chemistry-based biopolymeric hydrogels for regenerative medicine. Biomed. Mater. **16** (2021). https://doi.org/10.1088/1748-605X/abc0b3
80. Zhang, Y., Liu, S., Li, T., et al.: Cytocompatible and non-fouling zwitterionic hyaluronic acid-based hydrogels using thiol-ene "click" chemistry for cell encapsulation. Carbohydr. Polym. **236**, 116021 (2020). https://doi.org/10.1016/j.carbpol.2020.116021
81. Liu, K., Wei, S., Song, L., et al.: Conductive hydrogels—A novel material: recent advances and future perspectives. J. Agric. Food Chem. **68**, 7269–7280 (2020). https://doi.org/10.1021/acs.jafc.0c00642
82. Ballantyne, B., Myers, R.C.: The acute toxicity and primary irritancy of glutaraldehyde solutions. Vet. Hum. Toxicol. **43**, 193–202 (2001)
83. Lithium-ion batteries need to be greener and more ethical. Nature **595**, 7 (2021). https://doi.org/10.1038/d41586-021-01735-z
84. Liu, X., Steiger, C., Lin, S., et al.: Ingestible hydrogel device. Nat. Commun. **10**, 493 (2019). https://doi.org/10.1038/s41467-019-08355-2

85. Qian, C., Higashigaki, T., Asoh, T.-A., Uyama, H.: Anisotropic conductive hydrogels with high water content. ACS Appl. Mater. Interfaces **12**, 27518–27525 (2020). https://doi.org/10.1021/acsami.0c06853
86. Wang, Y., Zhang, L., Lu, A.: Transparent, antifreezing, ionic conductive cellulose hydrogel with stable sensitivity at subzero temperature. ACS Appl. Mater. Interfaces **11**, 41710–41716 (2019). https://doi.org/10.1021/acsami.9b15849
87. Liang, Y., Ye, L., Sun, X., et al.: Tough and stretchable dual ionically cross-linked hydrogel with high conductivity and fast recovery property for high-performance flexible sensors. ACS Appl. Mater. Interfaces **12**, 1577–1587 (2020). https://doi.org/10.1021/acsami.9b18796
88. Cui, Y., Chai, J., Du, H., et al.: Facile and reliable in situ polymerization of poly(Ethyl Cyanoacrylate)-based polymer electrolytes toward flexible lithium batteries. ACS Appl. Mater. Interfaces **9**, 8737–8741 (2017). https://doi.org/10.1021/acsami.6b16218
89. Crespilho, F.N., Sedenho, G.C., De Porcellinis, D., et al.: Non-corrosive, low-toxicity gel-based microbattery from organic and organometallic molecules. J. Mater. Chem. A (2019). https://doi.org/10.1039/C9TA08685D
90. Mark, M., Phillip, N., Alison, H., et al.: An ingestible bacterial-electronic system to monitor gastrointestinal health. Science **360**(80), 915–918 (2018). https://doi.org/10.1126/science.aas9315
91. Poosapati, A., Vadnala, S., Negrete, K., et al.: Rechargeable zinc-electrolytic manganese dioxide (EMD) battery with a flexible chitosan-alkaline electrolyte. ACS Appl. Energy Mater. **4**, 4248–4258 (2021). https://doi.org/10.1021/acsaem.1c00675

Biophotovoltaic: Fundamentals and Recent Developments

Gustavo P. M. K. Ciniciato

Abstract Although we have been increasing our understanding on how plants and autotroph microorganisms use photosynthesis to capture the energy from sunlight for producing their own energy, never have we used this knowledge to directly produce electricity until now, with the advent of biophotovoltaic. This chapter reviews recent research and development being conducted in this technology. It will begin with a brief discussion of microbial fuel cells, and similarities that led to the invention of biophotovoltaic. It will then be followed by a review regarding the influence on the nature of the microorganism, electrode materials, design of the device, and their effect on the power generation. The chapter will also cover the efforts some research groups have been undertaken to understand the phenomena behind electron transfer, and point out some of the major questions under debate concerning the functioning on biophotovoltaic. Finally, it will be presented a few examples of important steps some research groups have been conducting, and some future prospects regarding the use of biophotovoltaic. This chapter is thought as a simple guideline for bioelectrochemists who currently work with microbial fuel cells, and wish to start adventuring within biophotovoltaic.

1 Introduction

Life on Earth is directly linked to the Sun, its primary source of energy. The Sun, composed mostly of hydrogen under constant conversion to helium through nuclear fusion, radiates the leftover energy as electromagnetic waves to the space. The Earth located in ideal position, not too far and not too close to the Sun, is irradiated with enough energy and heat to maintain a high degree of organization where life can be sustained. Although primitive life form has an origin on geological features as source of energy, it is with the rise of photosynthesis that life acquired the necessary energy from the light to spread and evolve.

G. P. M. K. Ciniciato (✉)
Federal Institute of Education, Science and Technology of São Paulo, Av. Prof. Célso Ferreira da Silva, 1333, Avaré, SP 18707-150, Brazil
e-mail: gustavo.ciniciato@ifsp.edu.br

F. N. Crespilho (ed.), *Advances in Bioelectrochemistry Volume 4*,
https://doi.org/10.1007/978-3-030-99662-8_6

Humankind has learned how to use sunlight for a large number of application, from heating and drying to generating electricity. For instance, when production of electricity is considered, two main approaches are commonly used: (i) an array of heliostat mirrors can be used to concentrate sunlight into a boiler producing steam, and the acquired potential energy of the steam can be turned into kinetic energy to rotate a turbine, (ii) an assembly of photovoltaic cells (doped semiconductors) are mounted in a framework, where light promotes electron removal from a *n*-type semiconductor (semiconductor doped with an element containing excess electrons) which can be conducted through electric field toward a *p*-type semiconductor (semiconductor doped with an element containing fewer electrons), called a hole. So far, solar energy counts for only 2% of global electricity, but recently it was achieved as an increase from 30 to 840 TWh from 2010 to 2020 worldwide [1]. Thanks to advances and innovation in sciences, engineering, and manufacturing techniques, it is possible to decrease the cost of manufacturing and grid integration. To this day, silicon solar cells reached the module cost as lower as $1.00 W^{-1}, becoming even closer to the goal of $0.50 W^{-1} for the technology to become competitive [2].

Following the promise for a true clean and renewable source of energy that solar energy can provide, there is a novel technology being developed to produce electricity from sunlight. Therefore instead of artificially reproducing the ways how nature uses sunlight into a technology, it can be possible to get closer than ever to what nature developed as its best, photosynthesis. This technology is known as biophotovoltaic, a special type of biofuel cell.

2 General Aspects of Biofuel Cells

Fuel cells are electrochemical devices developed to convert chemical energy into electricity. It works through the oxidation of a fuel at the anode and electrochemical reduction of an oxidant at the cathode, where hydrogen and alcohols are commonly used as fuel, and oxygen from atmosphere is the most usual oxidant. Fuel cells require proper catalysts for the anodic and cathodic reactions, as well as an ion exchange membrane or an electrolyte between the electrodes to provide ionic conduction. This separator is required to maintain electroneutrality due to the charge imbalance resulted from electron transfer between anode and cathode.

Biofuel cells (BC) are a novel technology that has been developed for the generation of electricity using clean and renewable energy sources [3]. With similarity to fuel cells, a BC is composed of an anode containing proper biocatalysts where a substrate can be oxidized, generating electrons that can promote useful work (electricity) while moving toward a cathode containing proper biocatalyst where an oxidant, usually oxygen from air, can be reduced (Fig. 1). Conventional fuel cells make use of an ion exchange membrane to allow for the ionic conduction between biocathode and bioanode to maintain electroneutrality, but due to the nature of biocatalysts, it is common to make use of an electrolyte for the ionic conductivity. Ideally a buffer

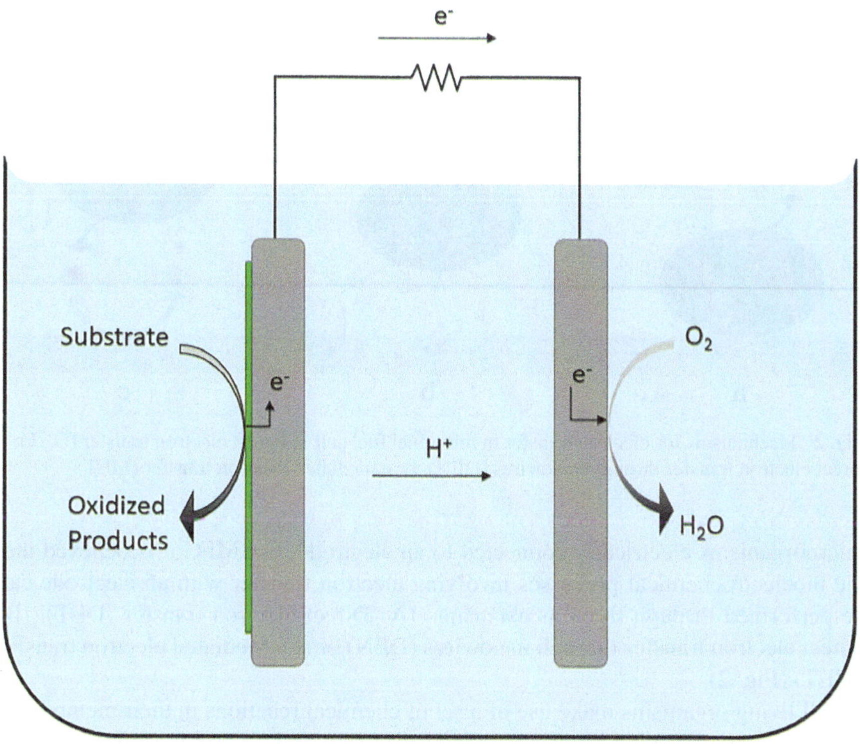

Fig. 1 Schematic of a conventional biofuel cell comprising a bioanode with a biocatalyst where a substrate can be oxidized and a cathode where a substrate (usually oxygen) can be reduced separated by an electrolyte, where ions can be conducted

solution is used due to the importance of keeping the optimum chemical environment for the biocatalysts.

Although some similarities to conventional fuel cells can be considered, biofuel cells benefit from the incorporation of biocatalysts directly to the reactions of interests, eliminating the use of expensive noble metal catalysts, which not only decrease the final cost of the electricity generation but also take advantages of using some intrinsic properties that biocatalysts particularly have such as: the use of low toxicity substrates coming from renewable sources, including glucose, lactate, alcohols and other; perform high activity under moderated conditions of pH and temperature; present high selectivity for the reactions of interest which avoid the formation of undesired products and the phenomena known as crossover of fuel, commonly observed in alcohol fuel cells [4]. In addition to that, the system can also be projected to use several biocatalysts in cascade of reactions to utilize the maximum energetic content in the primary fuel.

Biofuel cells are classified according to the source of the catalyst. It is called an enzymatic fuel cell (EFC) when the catalyst is an enzyme connected to an electrode or microbial fuel cell (MFC) when the catalyst involves a biofilm containing

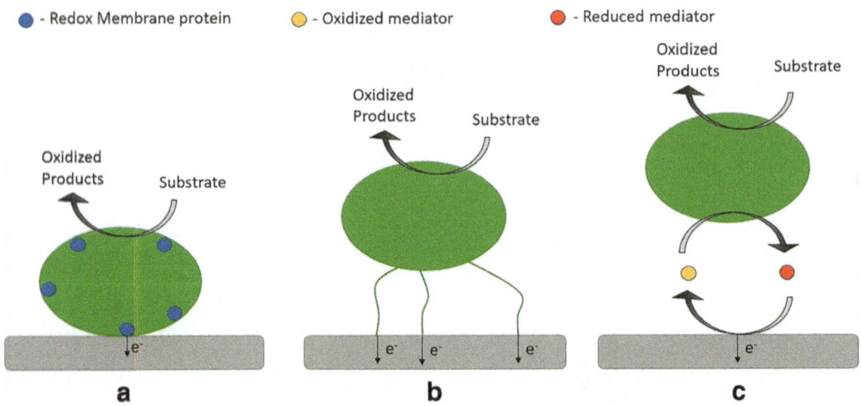

Fig. 2 Mechanisms for electron transfer in microbial fuel cell. **a** Direct electron transfer (DET), **b** direct electron transfer through nanowires (DEN), or **c** mediated electron transfer (MET)

microorganisms electrically connected to an electrode. For MFC it is believed that the bioelectrochemical processes involving electron transfer with an electrode can be performed through three mechanisms: (A) Direct electron transfer (DET), (B) Direct electron transfer through nanowires (DEN) or (C) Mediated electron transfer (MET) (Fig. 2).

All living organisms make use of a set of chemical reactions in their metabolism to produce the required energy to sustain life. Most of these reactions are redox in nature, where in the catabolic path a compound (food) is oxidized, and another compound is reduced through a set of electron transfer carried on enzymatically. Typically, aerobic microorganisms make use of oxygen as the final electron acceptor within its metabolic chain. Oxygen is ubiquitous, and it is energetically rich, behaving as a very strong oxidant ($E^0 = 1.229$ V vs SHE at 25 °C and pH 7.0). Oxygen can diffuse passively inside the cell, and participate in the metabolic reactions occurring in the cytosol as well as in the mitochondria. Anaerobic microorganisms on the other hand evolved using an electron acceptor other than oxygen for its respiration chain termination. Soluble compounds, such as humic substances, quinones, phenazines, or riboflavin, can function as exocellular electron mediators enhancing this type of anaerobic respiration [5]. Dissimilatory metal-reducing bacteria (DMRB), for example, use metal oxides outside their cell as final electron acceptors, making it part of their respiratory system [6]. Research in microbial bioelectrochemistry aim to make use of this type of microorganisms, since the respiratory chain contains part of its mechanism available externally to the cell wall that can be connected to an electrode. There are several examples of DMRB being electrochemically used that can be found in literature, such as reduction of Fe(III) [7], reduction of Mn(II) [8], and even reduction of U(VI) [9] as a final electron acceptor for its anaerobic respiration.

Microorganisms capable of DET are called exoelectrogenic. The DET occurs via a direct physical contact of the microorganism's cell membrane with the electrode, usually through redox active proteins, such as cytochromes or flavoproteins, located on the outer surface of the cell membrane. They allow for a possible transport of electrons between the biochemical processes occurring inside the cell with the external environment [10]. Due to this natural phenomenon of electron transfer, microbial attachment to the electrode, as well as electrical interactions with the electrode are crucial for an effective bioelectrochemical process. There are several reports in the literature of MFC being developed with microorganisms capable of DET including *Shewanella* spp., [11, 12], *Geobacter* spp. [13], *Rhodoferax ferrireducens* [14], *Thermincola potens* [15], *Clostridium* spp. [16] as some examples.

It was recently discovered that some microorganisms not only are capable to communicate electrochemically with the extracellular environment, but they can develop extensions (pili) from their cell wall to reach these metal oxides located in the environment, to improve cell adhesion, and to increase its contact surface area. These nanowires were proved to be metallic-like conductive containing terminal redox membrane proteins [17]. *Geobacter Sulfurreducens* for instance are capable to produce cytochrome rich nanowires with conductivity in the order of 5 mS cm^{-1}, and it is known as a mechanism it uses to reduce Fe(III) [18]. Another famous example of a microorganism capable of producing conductive nanowires is *Shewanella oneidensis* presenting similar behavior of electron transfer [19]. Bioelectrodes containing microorganisms that are able to transfer electrons through nanowires are considered to have mechanism of DEN, and a biofilm formed for them is called a bioelectroactive biofilm.

Due to weak interactions with an electrode, exoelectrogen biofilms running through DET or DEN usually present poor electron transfer, resulting in the production of low current density, with maximum values of current density in the magnitude of few mA cm^{-2} [20]. In order to improve electron transfer, one can make use of an electrochemical mediator to improve the bioelectrochemical mechanism with the electrode through MET. An electrochemical mediator can be defined as a reversible redox active component that can shuttle electrons from a biocatalyst to an electrode and vice versa. The mediator therefore has to be electrochemically compatible with the microorganism and with the electrode.

Although a mediator allows for an increase in current density by improving the number of biocatalysts interacting with the electrode, it has a price to add as an addition factor for the polarization activation, since the mediator has to react with the biocatalyst as well as with the electrode, it ends up providing a lower onset potential in comparison with the biocatalyst acting directly to the electrode. A mediator can be artificially added to the electrode chamber and, in some cases, a mediator can be naturally produced by the microorganisms following an extracellular electron transfer mechanism linked to its metabolism. Reports in literature have shown the use of artificial mediators such as neutral red, methylene blue, thionine, meldola's blue, and 2-hydroxy-1,4-naphthoquinone for the anode and ferricyanide as a mediator in the cathode replacing oxygen [21]. It was reported that some microorganisms can synthesize natural soluble mediators, such as flavins [22] and quinones [23] that help them to oxidize extracellular components.

In practice, one type of microorganism can perform more than one mechanism of electron transfer, and when considering a mixed community of microorganisms, it is expected to have a combination of different pathways for the electron transfer. That goes straight to the successful application of MFC since mixed cultures in general present better efficiency in comparison to pure cultures [24].

3 Biophotovoltaic

Basically, two types of energy sources are employed by microorganisms, light energy by phototrophic microorganisms and chemical energy by chemotrophic microorganisms. Biophotovoltaic (BPV), also known as photomicrobial fuel cells, photosynthetic microbial fuel cells, or microbial solar cells, function similarly to conventional microbial fuel cells. Therefore, instead of relying on the electrons involved in electron transport chain during metabolism of organic compounds in chemotrophs, BPV aims to make use of the high energy electrons provided by phototrophs, microorganisms capable of doing photosynthesis using the sunlight as a main source of energy, including eukaryotic microalgae and cyanobacteria (blue green algae) [25].

BPV may be considered a special case of MFC. Even though the bioanode is constructed with the deposition of biofilm containing microorganisms, the source of energy comes from the light. Conventional MFCs require supply for an organic substrate to be oxidized within the microorganism's metabolism, and mass transfer to the organisms generating extracellular electrons from the substrate may limit current generation. Mass transfer for BPV, on the other hand, is not a major limitation, since the source of electrons acquired in photosynthesis comes from the water (present in abundance), and light can be obtained from the sunlight [26]. In contrast to conventional photovoltaic systems, BPV systems based on whole cells may continue to produce some current in the dark due to heterotrophic metabolism related to the breakdown of stored carbon produced during the light period. In this case, BPV may work as a conventional MFC when light is not available [27].

The definition of photosynthesis has been debated over the years. One may consider an accurate interpretation as "a series of processes in which electromagnetic energy is converted to chemical energy used for biosynthesis of organic cell materials" [28]. Vascular plants, algae, and cyanobacteria are photosynthetic organisms capable of harvesting the incoming high energy photons from the light through an array of protein and chlorophyll molecules embedded in the thylakoid membrane, called light-harvesting antenna. This protein array transmits this energy to the reaction center through an exciton transfer (charge separation), creating a strong reducing agent to the photosystem II (PII), ($E'^{\circ} \sim 1.0$ V) which triggers an electron-transfer chain.

The high energy electrons extracted from reaction center PII are transferred to protein complexes containing metallo-organic cofactors such as pheophytin (Pheo), followed by a tightly protein-bond plastoquinone PQ_A, which then passes to another electron shuttle, a loose protein plastoquinone PQ_B. PQ_B, in its fully reduced quinol form, PQ_BH_2, reaches the cytochrome b6f complex (Cyt b6f), where electrons are transferred to it, and protons are pumped across the thylakoid membrane, producing the proton gradient required for ATP synthesis. Electrons are then transferred to photosystem I (PI) through a soluble protein plastocyanin (Plast).

Similarly to the PII, photons are harvested with light-harvesting antenna, and transmit this energy to PI, becoming a strong reducing agent ($E'^{\circ} \sim 1.5$ V), and triggering another electron-transfer chain. The electrons from Plast increase their energy, and are transferred to an acceptor, A_0. A_0 transfers these electrons to the next acceptor phylloquinone (A_1), which passes it to an iron-sulfur protein (Fe–S), and finally to another protein associated to the thylakoid membrane, ferredoxin (Fd). The terminal electron acceptor is a strong oxidant agent, $NADP^+$ ($E'^{\circ} = -0.324$ V) (Fig. 3).

There is a variety of compounds nature selected as electron source used for redox cycling, where oxygenic photosynthetic organisms use water as a source of electrons to regenerate PII. Due to evolution as well as adaptation to different environments, anoxygenic photosynthetic organisms utilize other electron donors instead of water, such as hydrogen, ferrous ion, or hydrogen sulfide [29]. Both the NADPH and ATP produced as a result from the light-dependent reactions in photosynthesis are used on the following dark reactions (known as Calvin cycle). The NADPH provides electrons while the ATP provides the required energy for the CO_2 fixation and carbohydrate production. The energy initially acquired from the photons reaching the light-harvesting antenna is converted to chemical energy and stored in the form of carbohydrates using the inorganic carbon from the atmosphere in the form of CO_2. The whole course of photosynthesis involves a series of electron transfer processes catalyzed by proteins and electron shuttles.

The goal of developing a biophotovoltaic aims to connect the electron transfer mechanism involved to photosynthesis within an electrode. The source of energy

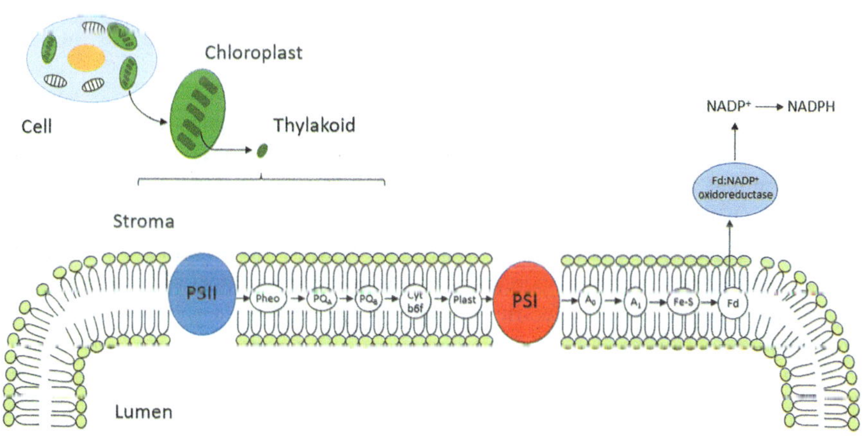

Fig. 3 Scheme presenting the electron pathway in eukaryotic photosynthetic organisms, and the electron shuttles involved with the electron transfer

comes from light, and the source of electrons is water, yielding oxygen in an ideal bioelectrochemical anodic oxidation. They have the capacity to generate electricity through bioelectrochemical reactions in two separated phases. When light source is available, this bioelectrochemical device works as a self-sustaining photosynthetic microbial fuel cell, converting the energy prevenient from the photosynthesis reactions into electrical energy. When light source is not available, this bioelectrochemical device works as a conventional microbial fuel cell (MFC), catalyzing the conversion of chemical energy from the carbohydrate reserve, which was formed during photosynthesis, into electrical energy.

Bioelectrochemistry requires electric contact between a biocatalyst with an electrode surface. The strategy behind developing a BPV involves connecting electrically some of these electron shuttles, and extracting directly or indirectly the high energy electrons involved in photosynthesis. An electrode reaction involves heterogeneous processes related to the adsorption/desorption of electroactive species with an electrode surface for the electron transfer to have a chance to be successful. By making a choice of a proper catalyst it is possible to decrease activation polarization by providing chemical pathways where energy barrier for the electron transfer is minimal. Electric and ionic conductivity are important aspects to be considered when electrode materials as well as membrane or electrolyte are chosen since they must not contribute heavily to the resistance polarization. The electrochemical device must be designed to improve mass transfer, and decrease mass transfer polarization. Therefore, the nature of the photosynthetic microorganism, the choice of materials, as well as the design of the electrochemical device play an important role in the BPV performance.

4 Manufacturing BPV

4.1 Production of the Algae Bioelectrode

Anam and co-workers presented an overview of photoanode materials for BPV, emphasizing that an ideal electrode must present transparency, electrical conductivity, chemical stability, biocompatibility, high surface area, low cost, and accessibility [30]. Traditionally bioelectrodes in microbial fuel cells are based on carbon materials. Some common electrode materials for MFC are graphite rod, graphite brush, carbon cloth, carbon paper, and carbon felt, among others. Carbon materials in general present excellent electrical conductivity, stability, large surface area, and strong biocompatibility [31].

Differently from conventional MFC, light is necessary as an energy source for BPV, and the choice for the electrode material has to take into account. Carbon materials in general are dark, and don't reflect light very well. The illumination of light being cast upon a carbon material can be absorbed, and the energy absorbed can be transferred through heat, increasing the temperature locally and possibly damaging the biofilm. The electrode material therefore not only has to be electrically conductive, but it should present a transparent or reflective surface. In addition to that, the reflexivity of the electrode surface will increase the time contact light will have with the biofilm, since light will pass through biofilm as an income electromagnetic wave and a second time upon reflection with the surface where biofilm is located. Some of the common choices for electrode materials to construct an anode for BPV are indium tin oxide (ITO), indium tin oxide-coated polyethylene terephthalate (ITO-PET) and fluor doped tin oxide (FTO), and other thin films of conducting materials such as reduced graphene oxide (rGO) deposited on glass slides [32].

A common way to promote a formation of the algae biofilm is by making use of simple adhesion of algae cell on the electrode surface (Fig. 4). A chosen volume of exponential phase culture is inoculated inside a sterile glass staining jar. The electrode is then placed inside the jar, and they are transferred to an incubator under illumination and temperature controlled according to the selected algae (A). Without shaking, the cells can be allowed to deposit on the surface of the electrode (B), and the electrode containing the biofilm can be removed from the jar (C) to be connected to the BPV [33]. Similarly, the solution containing algae can be concentrated by centrifugation (400 g, 10 min), suspended in fresh medium, and then pipetted on top of the electrode surface to be allowed to settle as a biofilm [27].

The nature of the microorganism, as well as the material of the electrode play an important role on how well the biofilm will develop on top of an electrode. Photoautotrophs biofilms are naturally found on illuminated surfaces where water is available, such as mountain streams, water treatment equipment or various natural and man-made structures in marine environment [34]. Microbial biofilm is formed through cell aggregation and production of extracellular polymeric substances (EPS). EPS is a complex mixture of macromolecules formed mostly by polysaccharides, proteins, nucleic acids, and lipids [35]. Although EPS is produced with the main role of promoting cell attachment to a surface, it may also enhance biofilm defense against environmental stress, improve cellular communication as well as increase surface area to reach nutrients. Algae biofilm strength depends largely on nitrogen and carbon availability. It was shown that enriching medium with nitrogen compounds

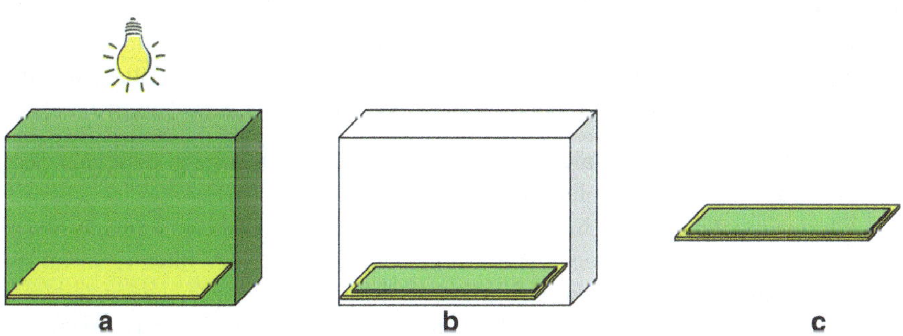

Fig. 4 Algae biofilm deposition on electrode surface to produce a BPV bioanode: **a** addition of the cell solution, **b** deposition on the electrode surface, and **c** bioelectrode containing the biofilm

may increase biofilm formation and adhesion ratio in consequence to increase on EPS production. Incorporation of glycine with concentration of 1 g L^{-1} is recommended as the optimal nitrogen concentration to improve algal biofilm formation [36].

It is imperative to keep track of the amount of algae matter is present in the biofilm deposited on the electrode surface, as well as changes that may occur during its use. Being formed of living organisms, the biofilm has a tendency to grow for as long as the nutrients and the environment are prosperous. The active biomass, as well as its thickness and distribution over the electrode strongly affect the electron transfer with the electrode [37]. The electrochemical surface area in contact with the biofilm is not an easy parameter to have under control due to the growth dynamic of the microorganisms. In fact, by increasing the surface area in a MFC does not necessarily result on the same increase on power production, since other factors such as reactor volume, as well as lack of enough electrochemical mediator when DET is not successful [38].

Optical density (OD) is commonly used as a tool to monitor photosynthetic cells in a photosynthetic biofilm because it is a simple and nondestructive technique. The OD of a photosynthetic suspension and biofilm is determined both by light scattering and by light absorption within the medium and the biofilm, being a function of concentration, size, and pigmentation of the cells. Under the pseudo steady-state conditions of exponential growth, size, biofilm thickness, and pigmentation remain constant, OD becomes directly proportional to the cell concentration [39]. Although a large number of pigments can be found, photosynthetic materials contain majority of chlorophyll and carotenoids. Chlorophyll, a porphyrin-ring molecule similar to that of hemoglobin but has a magnesium atom complexed to the ring instead of an iron atom. There are four types of chlorophyll, labeled a, b, c, and d, differentiating slightly in chemical structure, as well as in the region of the spectra they absorb. Each phylum differs in the amount and type of chlorophyll present, all algae contain chlorophyll a (Chl-a). Chl-a is insoluble in water and petroleum ether but soluble in alcohol, diethyl ether, benzene, and acetone. The pigment has two main absorption bands in vitro, one band with a maximum around red light region at 663 nm and the other at 430 nm [40].

The biofilm formed on top of the electrode surface must be removed for proper characterization. Since the interactions between the photosynthetic biofilm and the ITO surface are weak, biofilm can be easily removed using a pipette and washing it out with jets of water of fresh medium, and the suspended material is collected with a sterile beaker. Characterization can be performed following standard procedure [41]:

1. Two drops of a magnesium carbonate suspension (1 g in 100 mL distilled water) are added to the algae solution.
2. Photosynthetic cells are harvested by inverting the breaker into a second sterile flask using a filter paper 47 mm diameter Millipore AA filter or a 4.5 cm Whatman GF/C glass filter paper.
3. The filter is then placed in a 15 mL graduated centrifuge tube, add approximately 8 ml of 90% acetone and dissolve the filter by shaking the tube vigorously for 5–10 min.

4. Allow the pigments to be extracted by placing the tube in freezer (4 °C) in complete darkness for about 20 h.
5. Decant the clear supernatant liquid into a 10 mL 10 cm-path-length spectrophotometer cell.
6. Absorption of the supernatant containing the Chl-a extracted using 90% acetone is measured at 640 nm (OD_{630}), 645 $_{nm}$ (OD_{645}), and 665 $_{nm}$ (OD_{665}).

The Chl-a concentration is calculated using Eq. (1):

$$[Chl - a] = \frac{C_a \times V_a}{V_c} \tag{1}$$

where:

$[\mathrm{Chl} - a]$ is the chlorophyll a concentration in mg m^{-3};

V_a is the volume of acetone used for Chl − *a* extraction in mL;

V_c is the volume of the culture in mL;
C_a is the concentration factor for Chlorophyll *a*, calculated as follows [41]:

$$C_a = 11.6 \times OD_{665} - 1.31 \times OD_{645} - 0.14 \times OD_{630} \tag{2}$$

Similarly, Eq. (1) can be used to estimate concentrations of chlorophyll b (Chl-b) and chlorophyll c (Chl-c) using the concentration factor, and related optical densities as follows:

$$C_b = 20.7 \times OD_{645} - 4.34 \times OD_{665} - 4.42 \times OD_{630} \tag{3}$$

$$C_c = 55.0 \times OD_{630} - 4.64 \times OD_{665} - 16.3 \times OD_{645} \tag{4}$$

Another important parameter that can be used to monitor the biofilm status is the specific grown rate (μ), which is an indirect way of measuring the increase on the biofilm mass by considering an increase on the chlorophyll A content in consequence to the mass change during the logarithmic phase. The specific growth can be determined through measurements of OD_{650} over time, and calculated using the Eq. (5) [33, 42]:

$$\mu = \frac{ln(N_2 - N_1)}{t_2 - t_1} \tag{5}$$

where:

μ is the specific grown rate in h^{-1} or day^{-1};

N_2 is the OD650 measured at t_2;

N_1 is the OD650 measured at t_1;

t_1 and t_2 are the time periods considered within the exponential phase in hours or days.

The algal/cyanobacteria content that will be used to form the electrode biofilm usually is selected in the stationary phase. The mass concentration commonly used is in terms of dry weight (DW). A sample must be harvested using filter paper from a known volume V of the culture, and it is oven-dried at 100 °C for 24 h or until mass of the content is constant. The DW can be calculated using Eq. (6) [33]:

$$DW = \frac{w_d - w_0}{V} \tag{6}$$

where:

DW is the biofilm dry weight in mg L^{-1};

V is the volume of the culture in L;

w_d is the weight of the dried filter containing the algae biomass in mg;

w_0 is the weight of the blank filter in mg.

McCormick and co-workers studied the biofilm formation of two algae (*Chlorella vulgaris, Dunaliella tertiolecta*) and two cyanobacteria (*Synechocystis* sp. PCC 6803, *Synechococcus* sp. WH5701) strains on top of an ITO-PET electrode. By inoculating 150 nmol Chl-a of each strain for 8 days under low light (5 W m^{-2} and 12 h light/dark), they observed a formation of biofilm attached on ITO-PET with 82 ± 0.9 of the cells attached for *Synechococcus* sp. WH 5701, $77 \pm 4.1\%$ for *Chlorella vulgaris*, $74 \pm 10.5\%$ for *Dunaliella tertiolecta*, and $8.5 \pm 0.13\%$ for *Synechocystis* sp. PCC 6803. They observed the highest biofilm density 25.5 ± 1.0 pmol Chl-a mm^{-3} for *Chlorella vulgaris*, followed by 10.4 ± 0.6 pmol Chl-a mm^{-3} for *Synechococcus* sp. WH 5701. Authors observed a correlation between Chl-a content, biomass accumulation on the electrode, and power production from the anodic biofilm. In fact, *Synechococcus* sp. WH 5701 showed the highest power output in the light among the other photosynthetic microorganisms studied [27].

Modifying the ITO surface may affect how biofilm forms and in consequence affect the bioelectrode performance. By etching the ITO surface in lines with intervals of 2.5 and 1 mm using a diamond tipped pencil, it was possible to improve the biofilm coverage for *Chlorella* UMACC 313 from $90.50 \pm 7.58\%$ on smooth surface to $99.46 \pm 0.43\%$ on surface treated with 1 mm etching intervals. On the other hand, *Spirulina* UMACC 159 formed a biofilm with highest coverage, $80.70 \pm 1.86\%$ on the surface with 2.5 mm etching intervals [43].

4.2 Manufacturing the Biophotovoltaic

As any conventional electrochemical cell, biophotovoltaic requires an anode, a cathode, and a separator to allow ionic conductivity, and maintain electroneutrality in consequence to the charge separation caused by the electron transfer reactions between the electrodes. Difference arises from the fact that as light is the source of energy, the anode containing the biofilm catalyst must be designed to be open and be facing up, toward the sun. Cathode ideally must be designed to reduce molecular oxygen present in atmosphere due to high reduction potential, therefore an air-breathing cathode is desired.

Due to the intrinsic nature of photoautotroph microorganisms, the device must be designed to work under aqueous solution. Water not only serves as the source of electrons for photosynthesis, but it is also required to maintain cell structure through osmotic equilibrium, transport nutrients, and other soluble compounds to inside the cell, and in some cases can promote cell mobility. As long as substrates present in solution don't offer crossover, such as anolyte reacting on the cathode, and catholyte reaction at the anode, a membrane is not necessary, and the solution can work as an electrolyte.

By considering the works of McCormick [27], Bombelli [44], Periasamy [45], and Ng [46], one can observe a pattern for the device design. Although the device is made of different pieces of Perspex and Polydimethylsiloxane (PDMS) assembled together to form the body of the biophotovoltaic, the major structure with the main components is presented in Fig. 5.

Biophotovoltaic comprises structurally of a chamber where the solution containing the aqueous electrolyte as the media used to maintain the microorganisms

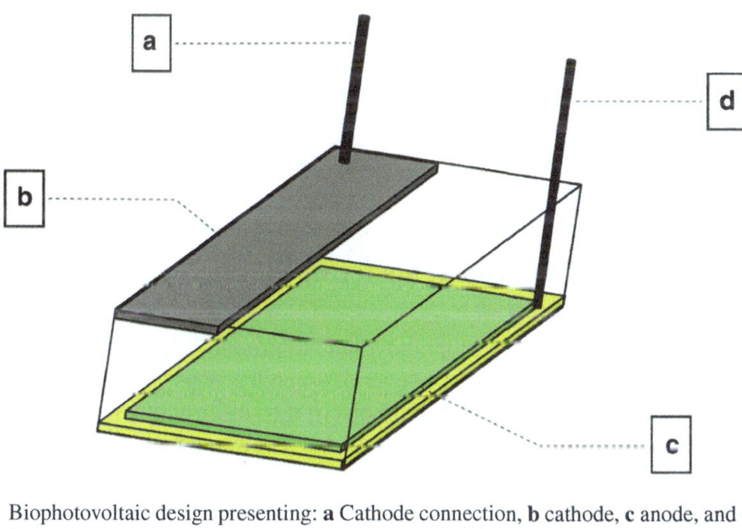

Fig. 5 Biophotovoltaic design presenting: **a** Cathode connection, **b** cathode, **c** anode, and **d** anode connection

is added (Fig. 5). The electric connection for the cathode (A) made of a conductive material such as stainless steel or carbon tape is connected to an air-breathing cathode (B) having hydrophilic side facing solution, and hydrophobic side facing air. Platinized carbon cloth or carbon paper is usually the choice for the cathode. The anode containing the autotroph biofilm (C) is usually placed at the bottom, facing up to receive the incoming light. The anode is electrically connected with the external circuit with a wire (D) that is usually isolated from the internal chamber, where the electrolyte containing water as the source of electrons, as well as nutrients to maintain the microorganisms healthy when it is necessary, according to the selected microorganisms.

The setup presented in Fig. 5 represents the whole electrochemical cell. By constructing the device with different pieces assembled together, it can be expanded to increase the size of anode and cathode, as well as increase the chamber volume. Larger electrodes are designed to provide higher current. Electron transfer within biofilm and electrode are usually poor, leading to low current density for the anode, but since most growth medium is prepared under neutral pH, it is expected for a cathode based on platinum to have low current as well. The proportion for the size of the electrodes is considered to balance current, since one of the electrodes may be limiting, but one must consider that since the cathode is located at the top, it may block the light from reaching the anode.

The volume of the camber is an important factor to be considered. First of all, since the device is supposed to be open to the environment, evaporation of the solution will eventually happen. Furthermore, due to the fact that water is being used as the source for the electron in photosynthesis, it is common to observe rapid decrease of the solution volume over time, so continuous replenishment of water is required. Care must be taken when considering scaling up a biophotovoltaic, since the growth medium usually does not present high conductivity, it behaves as a poor electrolyte. Therefore, distance between electrodes must be kept as low as possible, typically in the order of a few centimeters to avoid ohmic losses. A biophotovoltaic can be easily increased horizontally but not necessarily vertically. Additional salts can be added to the solution to improve the electrolyte conduction capacity, but the effect to the cell health due to the increased ionic strength must be considered.

5 Exploring the Electron Transfer in Biophotovotlaic

The proof of concept involving biophotovoltaic to produce electricity as an electrochemical device was done recently by several research groups, but so far very little is known regarding the mechanisms behind electron transfer photosynthetic microorganisms with an electrode. This is not restricted to autotroph microorganisms. Microbial fuel cells based on non-photosynthetic microorganisms present challenge for the understanding to this day. Several efforts are being made to elucidate mechanisms, and to understand what really is happening at the interface biofilm/electrode.

Bradley and co-workers modified *Synechocystis* creating mutants with respiratory terminal complexes removed to observe their effect within photosynthetic electron transfer using ferricyanide as an electron acceptor. They observed that with inactivation of terminal oxidase complexes, part of the electron flux was redirected to ferricyanide in the dark condition, resulting in increase on power production. While on low light conditions, all electrons generated by photolysis are used for carbon fixation, while under dark these electrons accumulate as reducing equivalents that are allowed to transfer to ferricyanide [47].

Pisciotta and co-workers studied the exoelectrogenic activity of 10 different genera of cyanobacteria, observing increase on voltage for all of them during light phase with decrease on dark phase for several days. An anode without cyanobacteria did not display the same response. They used chemical inhibitors for specific sites of the photosynthetic electron-transfer chain with bioelectrodes containing the cyanobacteria of genera *Nostoc* and *Lynbgya.* To probe the effect to the electron-transfer chain with the electrode process, 3-(3,4-dichlorophenyl)-1,1-dimethylurea (DCMU), and carbonyl cyanide m-chlorophenylhydrazone (CCCP), both known to prevent electron transfer between P680 (PSII), as well as phenylmercuric acetate (PMA), an inhibitor of P700 (PSI) were used (Fig. 6). They observed that CCCP, DCMU, as well as PMA reduced the electrogenic activity of illuminated bioelectrodes. DBMIB, which presents activity with cytochrome b6f increased considerably electrogenic activity. Their results suggest that the electron responsible for the electrogenic activity originates from P680 mediated biophotolysis of water, and PQ plays an important role in the electron transfer. This information was confirmed with a mediator tetramethyl-p-benzoquinone (duroquinone), which is known to compete with DCMU for binding to the PQ_B site on PS-II, and was able to restore electrogenic activity as a substitute to PQ_B after inhibition by DCMU [48].

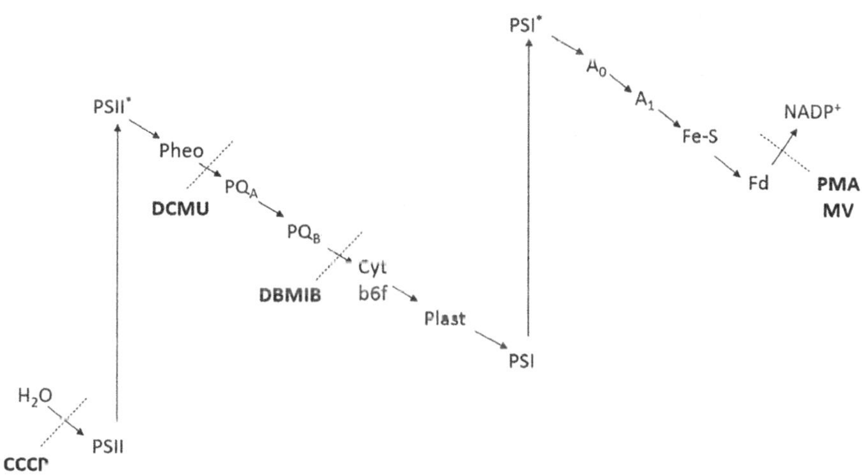

Fig. 6 Electron transfer pathway in photosynthesis and chemical inhibitors, CCCP, DCMU, DBMIB, PMA, and MV with their specific targets

In similar fashion, Bombelli and co-workers controlled a *Synechocystis* bioelectrode using inhibitors 2,5-dibromo-3-methyl-6-Isopropylbenzoquinone (DBMIB), known to accept electrons from PQ, and methyl viologen (MV), a competitor for the electrons from $NADP^+$. They observed that DCMU decreased power produced by *Synechocystis* bioelectrode by $63 \pm 17\%$, and blocked photosynthetic oxygen production when compared to the inhibitor-free control. They also observed that by using methyl viologen, a compound that can cross cell membranes, and compete with $NADP^+$ electrons involved in photosynthesis. In fact, little amount of MV decreased drastically the power response of BPV. It suggests that the electron collected by the electrode may leave the photosynthetic electron-transfer chain at the reductive side of PSI as Fd or NADPH [44].

Aiming on understanding the role of P700 on photocurrent generation, Cereda and co-workers created genetically modified *Synechocystis* sp. PCC6803 with limited functioning of P700. They observed that power production was strongly affected by modification, suggesting that electrons collected from electrode are directly related to electrons derived from water splitting by PSII [49].

Ciniciato and co-workers combined electrochemical techniques with pulse amplitude modulation fluorescence (PAM) to assess photosynthetic efficiency of the algae *Chlorella* sp. UMACC 313, and the cyanobacterium *Synechococcus elongatus* UMACC 105 biofilms on ITO electrodes with the effect of power production [50]. Four photosynthetic parameters were observed with the increase in electric current generation: the maximum quantum yield (MQY), the slope of the light-limited part of light curve, termed alpha (α), the light saturation coefficient (Ek), and the maximum relative electron transfer rate ($rETR_m$).

The MQY represents how efficiently the photosynthetic machinery responds to a pulse of light after complete adaptation in dark. The α is related to the light intensity required for the microorganism to reach their maximum photosynthetic activity. E_k represents the light intensity where photosynthetic rate is optimal. $rETR_m$ is the maximum electron transfer rate during photosynthesis. With an applied potential difference between cathode and anode moving from open circuit potential (OCP) going downward, it is observed an increase in current density for both studied microorganisms. The increase in current density is followed by a slightly decrease to the MQY, related to the weakening of the lipid bilayer with the rising of the electric field. On the other hand, the values of alpha and E_k increase with the increase of the cell potential, suggesting that photosynthetic efficiency may be enhanced with increased voltage. Finally, with current density being transferred to the electrode, it is observed a decrease on the $rETR_m$ for both bioelectrodes. The shape of the polarization curve shape matches the increase on current density with the decrease on $rETR_m$, suggesting that someway electrons from the photosynthetic pathway may be "leaking" to the electrode [50].

6 Recent Development in BPV

Biophotovoltaic is still in the level of research, and development. Several groups worldwide have been working on testing different microorganisms, developing new materials, modifying the electrode surface, or using chemicals such as mediators with the goal to obtain enough power to make this technology competitive to the current available ways of producing electricity.

Below is shown Table 1 presenting a selection of a few influential works hand selected by the author that have demonstrated improvement to the development of biophotovoltaic to this date.

Table 1 Recent influential works that contributed to BPV improvement

References	Short description
Firoozabadi et al. [51]	*Synechocystis* sp. PCC 6803 in BG-11 supplemented with NH_4Cl as anode with a platinized carbon cloth as air_cathode in single chamber BPV. NH_4Cl increased considerably the power output stimulating cell electrogenesis metabolic pathways reaching maximum power of 148.27 mW m^{-2}, one of the highest power produced by a BPV to this time
Thong et al. [52]	*Chlorella* sp. UMACC 313 on ITO modified with Cu and Cu-Ni metal organic framework as anode with platinized carbon paper as air–cathode separated by a Nafion-115 membrane. Synergies of Cu^{2+} and Ni^{2+}, reduced biofilm stress and improved current density generation as well as stability in comparison to bare ITO
Liu and Choi [53]	*Synechocystis* sp. PCC 6803 modified with intracellular gold nanoparticles enhanced the maximum power density by 33.6 times compared with the device without the nanoparticles in a paper based biophotovoltaic
Clifford et al. [54]	Pyocyanin was shown as a mediator that can harvest electrons directly from within the cell of *Synechocystis* sp. PCC 6803 to an ITO electrode with compatible potential to the photosynthetic reactions, and presenting high and stable current density
Karthikeyan et al. [55]	Ruthenium oxide (RuO_2)/tungsten oxide (WO_3) composite nanofibers presented as a good catalyst for the bioelectrodes containing *Chlorella vulgaris*, presenting a high and stable current under light
Bombelli et al. [56]	Microfluidic biophotovoltaic was constructed using soft-lithography using *Synechocystis* sp. PCC 6803 cells on InBiSn as bioanode, produced high power density with peak power in order of 100 mW m^{-2}
Ng et al. [57]	Use of Langmuir–Blodgett method allowed for the production of three-dimensional reduced graphene oxide transparent layers on ITO, increasing its efficiency to 119% in comparison to plain ITO
Thorne et al. [58]	Porous three-dimensional FTO coated TiO_2 ceramic electrode containing *Chlorela Vulgaris* promote dense colonies of algae, and increased power production to around 450 times in comparison to the electrode without FTO

7 Present and Future of BPV

Even though biophotovoltaic is still a technology under development, there has been a considerable effort to improve this technology, as well as to understand the bioelectrochemical phenomena happening in the interface biofilm/electrode for power generation. One can observe that the majority of publications on BPV comes from research being done by researchers at the University of Cambridge, UK, and the University of Malaya, Malaysia.

McCormick and co-workers elaborated an estimation of maximum power output that can be obtained from a cyanobacterium biophotovoltaic working on regions in the northern hemisphere, making a few considerations. They considered the light reactions of photosynthesis utilize around 50% of the available solar light. From these, approximately 90% is efficiently absorbed by cyanobacteria. They considered conservatively that for every 5 mols of photons absorbed, 2 mol of electrons for the reduction of one mole of $NADP^+$ were used. Due to cellular metabolic processes, 1/3 of the electrons generated by photosynthesis are lost, so only 2/3 are available. Finally, electrochemical conversion efficiencies in MFC typically range from 60 to 95%. Considering all these assumptions, McCormick and co-workers estimated an electrical current ranging from 3400 to 24,600 mA m^{-2}, with an average power density between 700 and 7700 mW m^{-2} [59].

Considering the literature, one can find publications showing power density around 100 mW m^{-2} in 2014 [56]. In 2021 a power output of around 150 mW m^{-2} was reached [51]. With the understanding in the mechanisms behind electron transfer between autotroph microorganisms, and electrodes, development of new materials, and even genetic engineering, it is possible to reach the estimated power density, and make this technology competitive.

In regards to intellectual property, one patent related to biophotovoltaic is active to this date. The Patent No.: US 10,665,866 B2, filed by University Malaya with the inventors: Periasamy Vengadesh, Siew Moi Phang, Fong Lee Ng, Mussoddiq Muhammad, Adrian C. Fisher, and Yunus Kamran, describing the fabrication, and integration of graphene electrodes with algal biofilm, and application to BPV.

References

1. Ritchie, H., Roser, M.: Renewable Energy (2020). https://ourworldindata.org/renewable-energy
2. Chu, S., Majumdar, A.: Opportunities and challenges for a sustainable energy future. **488**, 294–303 (2012). https://doi.org/10.1038/nature11475
3. Bullen, R.A., Arnot, T.C., Lakeman, J.B., Walsh, F.C.: Biofuel cells and their development. Biosens. Bioelectron. **21**, 2015–2045 (2006). https://doi.org/10.1016/j.bios.2006.01.030
4. Logan, B.E., Hamelers, B., Rozendal, R., Schröder, U., Keller, J., Freguia, S., Aelterman, P., Verstraete, V., Rabaey, K.: Microbial fuel cells: methodology and technology **40**, 5181–5192 (2006). https://doi.org/10.1021/es0605016

5. Stams, A.J.M., Bok, F.A.M., Plugge, C.M., Eekert, M.H.A.V., Dolfing, J., Schraa, G.: Exocellular electron transfer in anaerobic microbial communities. **8**, 371–382 (2006). https://doi.org/10.1111/j.1462-2920.2006.00989.x
6. Logan, B.: Exoelectrogenic bacteria that power microbial fuel cells. **7**, 375–381 (2009) https://doi.org/10.1038/nrmicro2113
7. Holmes, D.E., Bond, D.R., Lovley, D.R.: Electron transfer by *Desulfobulbus propionicus* to Fe(III) and graphite electrodes. **70**, 1234–1237 (2004). https://doi.org/10.1128/AEM.70.2.1234-1237.2004
8. Beliaev, A.S., Saffarini, D.A.: *Shewanella putrefaciens* mtrB encodes an outer membrane protein required for Fe(III) and Mn(IV) reduction. **180**, 6292–6297 (1998). https://doi.org/10.1128/JB.180.23.6292-6297.1998
9. Fredrickson, J.K., Zachara, J.M., Kennedy, D.W., Duff, M.C., Gorby, Y., Li, S.W., Krupka, K.M.: Reduction of U(VI) in goethite (α-FeOOH) suspensions by a dissimilatory metal-reducing bacterium. **64**, 3085–3098 (2000). https://doi.org/10.1016/S0016-7037(00)00397-5
10. Pankratova, G., Hederstedt, L., Gordon, L.: Extracellular electron transfer features of Gram-positive bacteria. **1076**, 32–47 (2019). https://doi.org/10.1016/j.aca.2019.05.007
11. Xiong, Y., Shi, L., Chen, B., Mayer, U., Lower, B.H., Londer, Y., Bose, S., Hochella, M.F., Fredrickson, J.K., Squier, T.C.: High-affinity binding and direct electron transfer to solid metals by the *Shewanella oneidensis* MR-1 outer membrane c-type cytochrome OmcA. **128**, 13978–13979 (2006). https://doi.org/10.1021/ja063526d
12. Kim, B.H., Kim, H.J., Hyun, M.S., Park, D.S.: Direct electrode reaction of Fe(III) reducing bacterium, *Shewanella putrefaciens*. **9**, 127–131 (1999). https://doi.org/10.1002/(SICI)1099-1514(199905/06)20:3<127::AID-OCA650>3.0.CO;2-I
13. Holmes, D.E., Chaudhuri, S.L., Nevin, K.P., Mehta, T., Methé, B.A., Liu, A., Ward, J.E., Woodard, T.L., Webster, J., Lovley, D.R.: Microarray and genetic analysis of electron transfer to electrodes in *Geobacter sulfurreducens*. **8**, 1805–1815 (2006). https://doi.org/10.1111/j.1462-2920.2006.01065.x
14. Chaudhuri, S.K., Lovley, D.R.: Electricity generation by direct oxidation of glucose in mediatorless microbial fuel cells. **21**, 1229–1232 (2003). https://doi.org/10.1038/nbt867
15. Wrighton, K.C., Thrash, J.C., Melnyk, R.A., Bigi, J.P., Byrne-Bailey, K.G., Remis, J.P., Schichnes, D., Auer, M., Chang, C.J., Coates, J.D.: Evidence for direct electron transfer by a gram-positive bacterium isolated from a microbial fuel cell. **77**, 7633–7639 (2011). https://doi.org/10.1128/AEM.05365-11
16. Choi, O., Kim, T., Woo, H.M., Um, Y.: Electricity-driven metabolic shift through direct electron uptake by electroactive heterotroph *Clostridiumpasteurianum*. **4**, 1–10 (2014). https://doi.org/10.1038/srep06961
17. Kumar, R. Singh, L. Wahid, Z.A. Din, M.F.M.: Exoelectrogens in microbial fuel cells towards bioelectricity generation: a review. **39**, 1048–1067 (2015). https://doi.org/10.1002/er.3305
18. Malvankar, N.S., Lovley, D.R.: Microbial nanowires: a new paradigm for biological electron transfer and bioelectronics. **5**, 1039–1046 (2012). https://doi.org/10.1002/cssc.201100733
19. Gorby, Y.A., Yanina, S., McLean, J.S., Rosso, K.M., Moyles, D., Dohnalkova, A., Beveridge, T.J., Chang, I.S., Kim, B.H., Kim, K.S., Culley, D.E., Reed, S.B., Romine, M.F., Saffarini, D.A., Hill, E.A., Shi, L., Elias, D.A., Kennedy, D.W., Pinchuk, G., Watanabe, K., Ishii, S., Logan, B., Nealson, K.H., Fredrickson, J.K.: Electrically conductive bacterial nanowires produced by *Shewanella oneidensis* strain MR-1 and other microorganisms. **103**, 11358–11363 (2006). https://doi.org/10.1073/pnas.0604517103
20. Schröder, U.: Anodic electron transfer mechanisms in microbial fuel cells and their energy efficiency. **9**, 2619–2629 (2007). https://doi.org/10.1039/B703627M
21. Ieropoulos, I.A., Greenman, J., Melhuish, C., Hart, J.: Comparative study of three types of microbial fuel cell. **37**, 238–245 (2005). https://doi.org/10.1016/j.enzmictec.2005.03.006
22. Marsili, E., Baron, D.B., Shikhare, I.D., Coursolle, D., Gralnick, J.A., Bond, D.R.: *Shewanella* secretes flavins that mediate extracellular electron transfer. **105**, 3968–3973 (2008). https://doi.org/10.1073/pnas.0710525105

23. Ramasamy, R.P., Gadhamshetty, V., Nadeau, L.J., Johnson, G.R.: Impedance spectroscopy as a tool for non-intrusive detection of extracellular mediators in microbial fuel cells. **104**, 882–891 (2009). https://doi.org/10.1002/bit.22469
24. Nevin, K.P., Richter, H., Covalla, S.F., Johnson, J.P., Woodard, T.L., Orloff, A.L., Jia, H., Zhang, M., Lovley, D.R.: Power output and columbic efficiencies from biofilms of *Geobacter sulfurreducens* comparable to mixed community microbial fuel cells. **10**, 2505–2514 (2008). https://doi.org/10.1111/j.1462-2920.2008.01675.x
25. Howe, C., Bombelli, P.: Electricity production by photosynthetic microorganisms. **4**, 2065–2069 (2020). https://doi.org/10.1016/j.joule.2020.09.003
26. Wey, L.T., Bombelli, P., Chen, X., Lawrence, J.M., Rabideau, C.M., Rowden, S.J.L., Zhang, J.Z., Howe, C.J.: The development of biophotovoltaic systems for power generation and biological analysis. **6**, 5375–5386 (2019). https://doi.org/10.1002/celc.201900997
27. McCormick, A.J, Bombelli, P., Scott, A.M., Philips, A.J., Smith, A.G., Fisher, A.C., Howe, C.J.: Photosynthetic biofilms in pure culture harness solar energy in a mediatorless bio-photovoltaic cell (BPV) system. **4**, 4699–4709 (2011). https://doi.org/10.1039/C1EE01965A
28. Gest, H.: History of the word photosynthesis and evolution of its definition. **73**, 7–10 (2002) https://doi.org/10.1023/A:1020419417954
29. Hohmann-Marriott, M.F., Blankenship, R.E.: Evolution of photosynthesis. **62**, 515–548 (2011). https://doi.org/10.1146/annurev-arplant-042110-103811
30. Anam, M., Gomes, H.I., Rivers, G., Gomes, R.L., Wildman, R.: Evaluation of photoanode materials used in biophotovoltaic systems for renewable energy generation. **5**, 4209–4232 (2021). https://doi.org/10.1039/D1SE00396H
31. Zhou, M., Chi, M., Luo, J., He, H., Jin, T.: An overview of electrode materials in microbial fuel cells. **196**, 4427–4435 (2011). https://doi.org/10.1016/j.jpowsour.2011.01.012
32. Jaafar, M.M., Ciniciato, G.P.M.K., Ibrahim, S.A., Phang, S.M., Yunus, K., Fisher, A.C., Iwamoto, M., Vengadesh, P.: Preparation of a three-dimensional reduced graphene oxide film by using the langmuir-blodgett method. **31**, 10426–10434 (2015). https://doi.org/10.1021/acs.langmuir.5b02708
33. Ng, F., Phang, S., Periasamy, V., Yunus, K., Fisher, A.C.: Evaluation of algal biofilms on indium tion oxide (ITO) for use in biophotovoltaic platforms based on photosynthetic performance. **9**, 1–13 (2014). https://doi.org/10.1371/journal.pone.0097643
34. Melo, L.F.; Bott, T.R.; Fletcher, M.; Capdveville, B.: Biofilms—Science and Technology, pp. 1–708. Springer, Dordrecht (1992). https://doi.org/10.1007/978-94-011-1824-8
35. Branda S.S., Vik, S., Friedman, L., Kolter, R.: Biofilm: the matrix revisited. **13**, 20–26 (2005). https://doi.org/10.1016/j.tim.2004.11.006
36. Shen, Y., Zhu, W., Chen, C., Nie, Y, Lin, X.: Biofilm formation in attached microalgal reactors. **39**, 1281–1288 (2016). https://doi.org/10.1007/s00449-016-1606-9
37. Zhang, L., Zhu, X., Li, J., Liao, Q., Ye, D.: Biofilm formation and electricity generation of a microbial fuel cell started up under different external resistances. **196**, 6029–6035 (2011). https://doi.org/10.1016/j.jpowsour.2011.04.013
38. Ghangrekar, M.M., Shinde, V.B.: Performance of membrane-less microbial fuel cell treating wastewater and effect of electrode distance and area on electricity production. **98**, 2879–2885 (2007). https://doi.org/10.1016/j.biortech.2006.09.050
39. Myers, J.: In: Buriew, J.S (ed) Algal culture from laboratory to pilot plant, vol. 3, p. 11 (1976). https://doi.org/10.1093/aibsbulletin/3.5.11
40. Lee, R.E.: Phycollogy Book, pp. 1–614. Cambridge University Press, Cambridge (1999)
41. Strickland, J.D.H., Parsons, T.R.: A Practical Handbook of Seawater Analysis. The Alger press Ltd., Ottawa (1972)
42. Krzemińska, I., Pawlik-Skowrońska, B., Trzcińska, M., Tys, J. In-fluence of photoperiods on the growth rate and biomass productivity of green microalgae. **37**, 735–714 (2014). https://doi.org/10.1007/s00449-013-1044-x
43. Ng, F.L., Phang, S.M., Periasamy, V., Yunus, K., Fisher, A.C.: Algae biofilm on indium tin oxide electrode for use in bi-ophotovoltaic platforms. **895**, 116–121 (2014). https://doi.org/10.4028/www.scientific.net/AMR.895.116

44. Bombelli, P., Bradley, R.W., Scott, A.M., Philips, A.J., McCormick, A.J., Cruz, S.M., Anderson, A., Yunus, K., Brendall, D.S., Cameron, P.J., Davies, J.M., Smith, A.G., Howe, C.J., Fisher, A.C.: Quantitative analysis of the factors limiting solar power transduction by Synechocystis sp. PCC 6803 in biological photovoltaic devices. **4**, 4690–4698 (2011). https://doi.org/10.1039/C1EE02531G
45. Periasamy, V., Phang, S., Ng, F., Muhammad, M., Fisher, A.C., Yunus, K.: Method of fabricating graphene-based/algal biofilm electrode for application in a biophotovoltaic device. US Patent 10,665,866 (2020)
46. Ng, F., Phang, S., Lan, B.L., Kalavally, V., Thong, C., Chong, K., Periasamy, V., Chandrasekaran, K., Kumar, G.G., Yunus, K., Fisher, A.C.: Optimised spectral effects of programmable LED arrays (PLA)s on bioelectricity generation from algal-biophotovoltaic devices. **10**, 10105 (2020). https://doi.org/10.1038/s41598-020-72823-9
47. Bradley, R.W., Bombelli, P., Lea-Smith, D.J., Howe, C.J.: Terminal oxidase mutants of the cyanobacterium *Synechocystis* sp. PCC 6803 show increased electrogenic activity in biological photo-voltaic systems. **15**, 13611–13618 (2013). https://doi.org/10.1039/C3CP52438H
48. Pisciotta, J.M., Zou, Y., Baskakov, I.V.: Light-dependent electrogenic activity of cyanobacteria. **5**, 1–10 (2010). https://doi.org/10.1371/journal.pone.0010821
49. Cereda, A., Hitchcock, A., Symes, M.D., Cronin, L., Bibby, T.S., Jones, A.K.: A Bioelectrochemical approach to characterize extracellular electron Transfer by *Synechocystis* sp. PCC6803. **9**, 1–8 (2014). https://doi.org/10.1371/journal.pone.0091484
50. Ciniciato, G.P.M.K., Ng, F., Phang, S., Jaafar, M.M., Fisher, A.C., Yunus, K., Periasamy, V.: Investigating the association between photosynthetic efficiency and generation of biophotoelectricity in autotrophic microbial fuel cells. **6**, 1–10 (2016). https://doi.org/10.1038/srep31193
51. Firoozabadi, H., Mardanpour, M.M., Motamedian, E.: A system-oriented strategy to enhance electron production of *Synechocystis* sp. PCC6803 in bio-photovoltaic devices: experimental and modeling insights. **11**, 12294 (2021). https://doi.org/10.1038/s41598-021-91906-9
52. Thong, C., Priyanga, N., Ng, F., Pappathi, M., Periasamy, V., Phang, S., Kumar, G.G.: Metal organic frameworks (MOFs) as potential anode materials for improving power generation from algal biophotovoltaic (BPV) platforms. In press (2021). https://doi.org/10.1016/j.cattod.2021.07.020
53. Liu, L., Choi, S.: Enhanced biophotoelectricity generation in cyanobacterial biophotovoltaics with intracellularly biosynthesized gold nanoparticles. **506**, 1–8 (2021). https://doi.org/10.1016/j.jpowsour.2021.230251
54. Clifford, E.R., Bradley, R.W., Wey, L.T., Lawrence, J.M., Chen, X., Howe, C.J., Zhang, J.Z.: Phenazines as model low-midpoint potential electron shuttles for photosynthetic bioelectrochemical systems. **12**, 3328–3338 (2021). https://pubs.rsc.org/en/content/articlehtml/2021/sc/d0sc05655c
55. Karthikeyan, C., Kumar, T.R., Pannipara, M., Al0Sehemi, A.G., Senthilkumar, N., Angelaalincy, M.J., Phang, S., Periasamy, V., Kumar, G.G.: Ruthenium oxide/tungsten oxide composite nanofibers as anode catalysts for the green energy generation of *Chlorella vulgaris* mediated biophotovoltaic cells. **38**, 1–11 (2019). https://doi.org/10.1002/ep.13262
56. Bombelli, P., Müller, T., Herling, T.W., Howe, C.J., Knowles, T.P.K.: A high power-density, mediator-free, microfluidic biophotovoltaic device for cyanobacterial cells. **5**, 1401299 (2014). https://doi.org/10.1002/aenm.201401299
57. Ng, F., Jaafar, M.M., Phang, S., Chan, Z., Salleh, N.A., Azmi, S.Z., Yunus, K., Fisher, A., Periasamy, V.: Reduced graphene oxide anodes for potential application in algae biophotovoltaic platforms. **4**, 1-7 (2014). https://doi.org/10.1038/srep07562
58. Thorne, R., Hu, H., Schneider, K., Bombelli, P., Fisher, A., Peter, L.M., Dent, A., Cameron, P.J.: Porous ceramic anode materials for photo-microbial fuel cells. **21**, 18055–18060 (2011). https://doi.org/10.1039/C1JM13058G
59. McCormick, A.J., Bombelli, P., Bradley, R.W., Thorne, R., Wenzel, T., Howe, C.J.: Biophotovoltaics: oxygenic photosynthetic organisms in the world of bioelectrochemical systems. **8**, 1092–1109 (2015). https://doi.org/10.1039/C4EE03875D

Organic Semiconductors as Support Material for Electrochemical Biorecognition: (ii) Approaches and Strategies for the Design of High-Performance Devices

Nathália Galdino, Lara Fernandes Loguercio, Luiza de Mattos Manica, Carolina Ferreira de Matos, and Jacqueline Ferreira Leite Santos

Abstract The outstanding properties of organic semiconductors arouse interest in their application as electrochemical biosensors and their performance strongly depends on functionalization and detection strategies used. Therefore, this chapter presents some strategies for electrochemical biosensing, including different approaches for biomolecules immobilization and biorecognition by using several methodologies for detection and reading out. Finally, some strategies for designing miniaturized high-performance devices are described, such as the use of machine learning and screen-printed electrodes.

Keywords Organic semiconductor · Biomolecules immobilization · Electroanalytical approaches · Machine learning · Screen-printed electrodes

1 Electrochemical Biorecognition

When the biosensor is immersed into a solution containing the target molecules, specific bindings between recognition element previously immobilized in/on the electrode and the target molecule alter the electrical properties [5, 29], leading to changes that electrical read-out techniques can follow, such as:

(a) Amperometry (potentiostatic): a set potential is applied during a period while the electrical current is recorded. This technique is interesting to monitor a redox reaction that occurs at the set potential, for example, the oxidation of hydrogen peroxide generated by the glucose oxidase enzyme (GO_x).

N. Galdino · L. de Mattos Manica · J. F. L. Santos (✉)
Laboratory of Applied Materials and Interfaces/Institute of Chemistry, Federal University of Rio Grande do Sul, Porto Alegre, RS, Brazil
e-mail: jacqueline.ferreira@ufrgs.br

L. F. Loguercio
Northeast Center for Strategic Technologies, Recife, PE, Brazil

C. F. de Matos
Federal University of Pampa, Bagé, Brazil

F. N. Crespilho (ed.), *Advances in Bioelectrochemistry Volume 4*,
https://doi.org/10.1007/978-3-030-99662-8_7

(b) Potentiometry (galvanostatic): when a measurable potential or charge accumulation is produced. This technique is widely used to monitor changes in pH during enzymatic catalysis. For example, during the urea catalysis by the urease enzyme.
(c) Potentiometry (potentiodynamic): this technique allows obtaining in detail the detection by registering the potential and current related to the redox process. It is widely used in enzymatic biosensors, but not exclusively, allowing to monitor, for example, biorecognition events in aptasensor and immunosensors. For instance, cyclic voltammetry (CV) was used for DNA probe as the target sequence hybridization altered the electron flow through the CP backbone [5].
(d) Conductometry: when the detection alters the electrical conductivity or resistivity of the organic semiconductor biosensor. For example, an enzymatic biosensor produces hydrogen peroxide, which oxidizes the conducting polymers (CPs), changing its conductivity.
(e) Impedancemetry: an extremely sensitive technique based on changes at the electrode interface, allowing to monitor, for instance, the charge transfer resistance (Rct) and capacitance of the conductors and/or semiconductors. These changes are observed in CPs and carbon-based-modified electrodes after immobilizing the biorecognition element and detecting the target molecule, such as DNA, antibodies, specific binding proteins, and whole cells.

Other types of electrochemical detection techniques are available in the literature, such as by field-effect (which uses transistor technology to measure current because of a potentiometric effect at a gate electrode) [30, 34]. The choice for a suitable electroanalytical tool strongly depends on the characteristics of the biorecognition element. Hereafter some examples to detail the electrochemical biorecognition are presented.

1.1 Enzymes for Electrochemical Biorecognition

Enzymes are certainly the biomolecules more used to assemble electrochemical biosensors. They are versatile, allowing the determination of the analyte in diverse ways, such as by measuring:

(i) consumption of cosubstrates (e.g., O_2, H_2O_2);
(ii) formation of electroactive species (e.g., H_2O_2, Fe^{2+}, Fe^{3+}, ferrocyanide);
(iii) evolution of the biocatalyst redox state (e.g., catalase and peroxidase that contain Fe^{2+}, Fe^{3+} in their chemical structure);
(iv) changes in the electron flow rate straight from the enzyme active site to the transductor.

The history of electrochemical biosensors reports the importance of the enzymes in the evolution of these devices, being the GOx a protagonist, allowing the development of large equipments and miniaturized devices. One of the main interests of using

GO_X is related to the glucose levels in blood samples, as diabetes mellitus is a worldwide public health problem. This disease causes metabolic disorder from insulin deficiency and hyperglycemia, reflected by higher/lower blood glucose concentrations than the normal range of 80–120 mg dL^{-1} (4.4–6.6 mmol L^{-1}) [39]. Therefore, early diagnosis and treatment of diabetes mellitus require close monitoring of blood glucose levels.

Enzymatic electrodes based on GOx become a strong candidate for the replacement of continuous glucose monitoring methods. The GOx is produced naturally in some fungi and insects and possesses a flavin adenine dinucleotide cofactor (FAD) protein group in the active site [42]. Electrochemical enzymatic glucose biosensors can be divided into three different generations depending on the electrochemical communication between the enzyme and the electrode. Summarizing:

(i) The first generation involves the oxygen consumption in solution for the subsequent generation of hydrogen peroxide at high redox potential. After the oxygen consumption (cosubstrate), the product from the catalysis goes toward the transducer leading to an electrical response. This technology was used to produce the first biosensor for glucose in the market. The disadvantages include the use of platinum in the assembling (increasing the price of the device); the high potential to achieve high selectivity; the dependence on the solubility of oxygen in biological fluids; and the susceptibility to the enzyme inactivation due to the production of hydrogen peroxide [30].

(ii) The second generation uses electrochemical mediators (such as ferrocene, ferrocyanide, quinine, and methylene blue) to shuttle the electrons from the enzyme redox centers to the surface of the electrode. The presence of the mediator overcomes the limitation of low pressure of oxygen in the first generation. Additionally, the low redox potential of the mediator may decrease interferences from electroactive species such as uric acid and ascorbic acid. The main disadvantages concerning this generation are the competition between the mediator and dissolved oxygen; the fake and imprecise results from interferences of other electroactive species; the small size; and high diffusibility of the mediator, which may result in losses by leaving the region between the enzyme and the electrode [30].

(iii) Finally, in the third generation, the electron is transferred straightly from the enzyme active site to the electrode, applying a low operating potential close to the redox potential of the enzyme [23]. This configuration allows to improve the efficiency and response time of the biosensor since the only barrier for the electron flux is the globular enzyme structure. Some strategies to improve the electron transfer rate include modifying the electrode surface with carbon nanotubes (CNTs) or polypyrrole (PPy) to enable a suitable enzyme orientation [14, 28, 43].

There are plenty of works in the literature reporting the development of enzymatic electrochemical biosensors using organic semiconductors. For example, recently, a glucose biosensor was developed using nanocomposites of polyaniline (PANI) on TiO_2 nanoparticles (PANI-TiO_2 NPs) to overcome the limitations and performance of the pristine PANI or TiO_2 NPs [20]. According to Tang et al. [37], the synergistic effect between PANI and nanometer-sized TiO_2 not only provide a large specific surface area, good biocompatibility, and more active center sites for GOx but also helps to decrease the steric hindrance of the enzyme active centers, resulting in direct electron transfer from the enzyme to the electrode surface [37].

PANI-TiO_2 NPs were obtained by depositing an ultra-thin coating of PANI on the TiO_2 NPs surface by vapor phase polymerization [20]. The biosensors were fabricated by mixing PANI-TiO_2 NPs dispersed in 0.5% chitosan solution prepared in 1% acetic acid with GOx in PBS solution and then adding the resulting mixture on the dried surface of a glassy carbon electrode (GCE). Nafion solution was used as a binder for the preparation of the electrodes. Characterization with CV and electrochemical impedance spectroscopy showed a decrease of the Rct with the presence of PANI; on the other hand, the Rct increases after GO_x immobilization, as this enzyme acts as an electrical insulator.

The biosensor activity was studied with CV in both PBS purged with nitrogen and standard PBS [20]. There were two almost symmetrical redox peaks in the first case, indicating the direct electron transfer between the enzyme and the electrode surface (see Eq. 1). However, in the standard PBS, there were asymmetrical redox peaks because of the oxidation of the reduced enzyme by the dissolved oxygen (see Eq. 2), increasing the reduction peak current.

$$GO_X(FAD) + 2e^- + 2H^+ \leftrightarrow GO_X(FADH_2) \quad (1)$$

$$GO_X(FADH_2) + O_2 \rightarrow GO_X(FAD) + H_2O_2 \quad (2)$$

The biosensor performance in PBS purged with N_2 was evaluated in the presence of glucose. It was observed that the peak current reduction is strongly dependent on the glucose concentration. The enzyme reduction occurs during the glucose oxidation to gluconolactone (as shown in Eq. 3). The electrochemical impedance spectroscopy also indicated that the Rct increases with the addition of glucose. The active centers started to saturate with glucose concentration over 200 $\mu mol\ L^{-1}$ and observed a nonlinear behavior. The biosensor presented a good sensibility and limit detection of 5.33 $\mu mol\ L^{-1}$.

$$GO_X + \text{Glucose} \rightarrow GO_X(FADH_2) + \text{Gluconolactone} \quad (3)$$

Although not as popular as GOx, the enzyme acetylcholinesterase (AChE) has been extensively used as a biorecognition element on organic semiconductor support materials. One motivation for its application in the design of electrochemical biosensors is the uncontrolled use of pesticides, such as organophosphates and carbamates, which have high toxicity, leading to health problems for the worldwide population. These chemicals might contaminate, for example water, air, fruits, vegetables, fish, meat, milk, and eggs. Their toxicity is based on the inhibition of the enzyme AChE that is essential for the central nervous system of humans and insects. AChE function is to catalyze the hydrolysis of the neurotransmitter acetylcholine (ATCL) in thiocholine (TCh) and acetic acid. However, in the presence of pesticides, the nucleophilic group present in the active site of the enzyme binds covalently to the oxygen (carbamates) or phosphorus (organophosphates) inhibiting the biocatalytic activity of AChE. This inhibition results in the accumulation of ATCL, causing respiratory disorders, muscular problems, and dysfunctions of the myocardium, leading to death [27]. TCh is an electroactive product that can be oxidized [3, 33], as represented in Eq. 4.

$$\text{TCh} \rightarrow \text{dithiobischoline} + 2\text{H}^+ + 2\text{e}^- \quad (4)$$

Recently, Loguercio et al. reported the fabrication of biosensors based on PPy films doped with indigo carmine (IC), sodium dodecyl sulfate (DS), and Au NPs (PPy-IC-DS-Au NP) by in situ electrochemical polymerization onto indium tin oxide coated glass (ITO) for the detection of carbaryl pesticide using AChE (from *electric eel*) [19]. For the immobilization, a solution of AChE was drop cast on the PPy-IC-DS-Au NP film. They unveiled the dependence between the biosensor performance and the support material properties. By controlling the DS concentration, the authors increased the density of positive charges in the polymer surface, therefore the wettability, improving the electrostatic interaction with the enzyme above its isoelectric point (pI). Consequently, an expressive amount of active AChE was immobilized on the surface of the PPy films with their active site close to the interface (similarly to AChE from *electric ray*, as previously reported in a theoretical study [47]), enabling a fast direct electron transfer and high affinity to the pesticide. As the analytical curve followed the Langmuir isotherm model, the authors obtained the binding constant for the carbaryl pesticide toward the adsorption sites of the enzyme, reaching 7.39 $\times 10^8$. As a result, the biosensor presented high sensitivity, a low detection limit (0.033 ng cm^2 mL^{-1}), and demonstrated satisfying performance in tap water toward the recovery of carbaryl pesticide.

The CNTs are also known for facilitating the electron transfer from the enzyme to the transducer, improving the catalytic activity. Li et al. [15] had developed a xylose dehydrogenase-displayed bacteria combined with Nafion/multiwalled carbon nanotubes (MWNTs) composite as an amperometric biosensor to xyloses determination. This analyte is present in the food industry as sweetness and in medicine and industry processes since it might be a softener, surfactant, plasticizer, and the main hydrolysis product of cellulose. To avoid purification steps and destabilization of free enzymes, the xylose dehydrogenase was expressed on bacteria surface prepared from ice nucleation protein from *Pseudomonas borealis* DL7. In this case, the xylose detection was indirect. The NADH cofactor oxidation process at high positive potential transfers electrons to the electrode surface. The presence of MWNTs decreased the overpotential for NADH oxidation from ~0.7 to ~0.5 V due to the high local density of electronic states in MWNTs. The peak current and consequently the sensitivity are significantly higher when the electrode is modified with the nanomaterial, suggesting the NADH electrocatalytic oxidation by MWNTs [15].

1.2 Antibodies for Electrochemical Biorecognition

The electrochemical biosensors that use antibodies for the biorecognition with antigens are usually called electrochemical immunosensors and are applied in different areas, such as the medical, food industry, and environment. An interesting point of these devices is the universality of antibodies. Therefore, after demonstrating the proof of concept of an electrochemical immunosensor, changing the antibody in the assembling allows adapting the configuration for different purposes.

The research in this field focuses on improving their sensitivity as they have the potential to be miniaturized in Point of Care (PoC) devices with modern microelectronics that facilitates multiplexing and the reduction of the overall size. Some strategies consist of employing organic semiconductor nanomaterials. In addition, the detection may be signalized with electrochemical mediators or with enzyme-labeled molecules in a sandwich-type of immunoassay to improve the read-out [11]. Following are the description of some reports where organic semiconductors were used as support materials to assemble the electrochemical immunosensor.

The sensitive and selective detection of alpha-fetoprotein antibody (AFP-ab) plays a significant role in the early diagnosis and treatment of several types of cancer as AFP levels above 20 ng mL^{-1} are found in the human serum of patients [18, 44]. [8] developed an immunosensor of AFP using a poly (3,4-ethylenedioxythiophene) (PEDOT) doped with polyethylene glycol (PEG) terminated with thiol groups to interact with Au NPs. The composite was functionalized with AFP-ab by incubating the electrode in phosphate buffer solution with AFP-ab. Electrochemical impedance spectroscopy measurements were employed to evaluate the immunosensor performance. The schematic illustration for the immunosensor fabrication is summarized in Fig. 1. The AFP-ab is positively charged in the solution (pI = 8.0), attracting negative redox probes near the electrode surface, decreasing the Rct due to this enhancement of the charge transfer. Similarly, after the antigen detection, the Rct decreased. The immunosensor retained the AFP-ab high binding affinity, exhibiting favorable selectivity, high sensitivity, ultra-low detection limit (0.0003 fg mL^{-1}), demonstrating satisfying performance for the assay of AFP in human serum samples [8].

Graphene has been used as a scaffold to anchor specific biomolecules to detect analytes with high specificity. For example, a PoC continuous graphene-based device was developed by Salem et al. [31] to quantify fibrinogen, an indicator of trauma-induced coagulopathy in unprocessed whole blood samples. Fibrinogen-specific antibody 12HC was covalently attached to aminated-graphene surface through activation of reactive sites of graphene with succinic acid and EDC in an aqueous solution. The fibrinogen biosensor was prepared by dropping a 12HC-graphene solution on the working electrode of graphitic screen-printed electrode (SPE). An extra layer of chitosan solution hydrogel is essential to maintain the graphene layer onto the electrode surface. The miniature three-electrode biosensor was used to determine fibrinogen using electrochemical impedance spectroscopy and ferro/ferricyanide as the redox pair in PBS, plasminogen, or human serum albumin background, human serum, and unprocessed whole blood. The charge transfer process from the redox

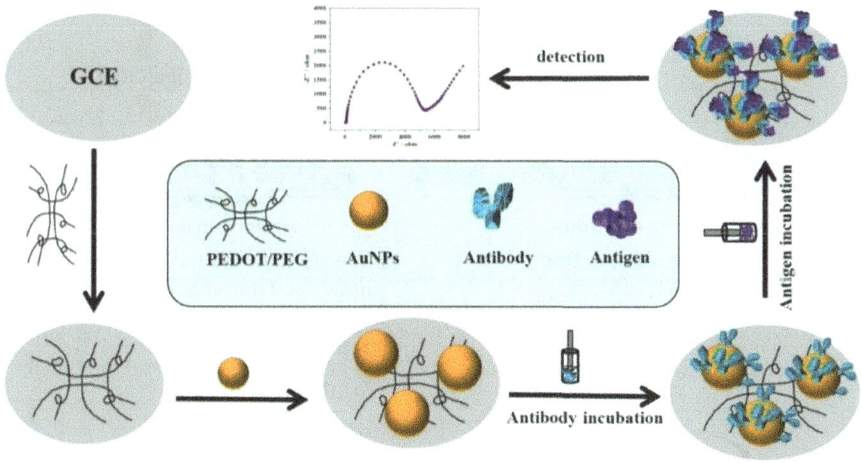

Fig. 1 Schematic illustration of the fabrication process of the AFP biosensor [8]. Copyright 2016 (Reprinted with permission from Elsevier Ltd.)

probe reaction to the electrode progressively hinders with the increase in fibrinogen concentration, which occupies the antibody sites on the graphene surface (Fig. 2) [31].

The configuration of an electrochemical immunosensor can be applied for the biorecognition of different analytes, including viruses, which have remarkable importance to prevent and control epidemics. Huang et al. [11] developed a sandwich-type immunosensor for avian influenza virus (AIV) H7 detection using linear sweep voltammetry. The graphene/chitosan/NPs nanocomposite was employed for both immobilization of monoclonal antibodies (MAbs) on Au NPs and as a tracer to label polyclonal antibodies (PAbs) along with silver nanoparticles (Ag NPs) activated with 1-ethyl-3-(3-dimethylaminopropyl) carbodiimide/N-hydroxysuccinimide (EDC/NHS). Figure 3 presents the sandwich immunoassay, where the AIV H7 is trapped in between the PAb-Ag NPs-graphene-chitosan and MAbs-Au NPs-graphene-chitosan layers after 30 min incubating. The immunosensor presented a linear response from 1.6 pg mL^{-1} to 16 ng mL^{-1} of AIV H7 concentration. The authors highlight the importance of graphene in the design of the electrochemical biosensor once the high surface-to-volume ratio allowed the immobilization of multiple Ag NPs, improving the response signal [11].

Also, a combination of immuno and enzymatic responses can be integrated into a single device. Torrente-Rodríguez et al. [38] developed a portable and wireless graphene-based electrochemical platform to diagnose COVID-19 in serum and saliva. The platform is expected to favor the diagnoses of viral infection, immune response, and disease severity by providing information on SARS-CoV-2 nucleocapsid protein (NP), specific immunoglobulins against SARS-CoV-2 spike protein (S1-IgM and S1-IgG), and inflammatory biomarker C-reactive protein (CRP) concentrations, respectively.

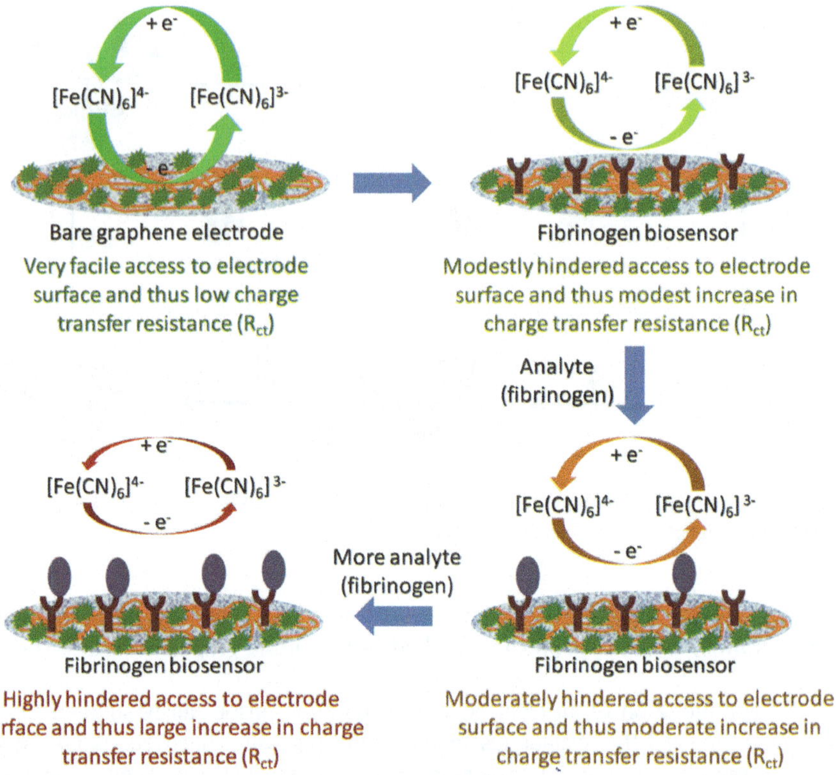

Fig. 2 Principle of fibrinogen detection via EIS spectroscopy using antibody covalently attached on the aminated-graphene surface electrode [31]. Copyright 2016 (Reprinted with permission from Elsevier Ltd.)

Fig. 3 Graphene sandwich-type immunosensor assemble; graphene/chitosan modification with (a) Au NPs for Mab immobilization and (b) Ag NPs activated with EDC/NHS for PAbs, and (c) biosensor detection of AIV H7 [11]. Copyright 2016 (Reprinted with permission from Elsevier Ltd.)

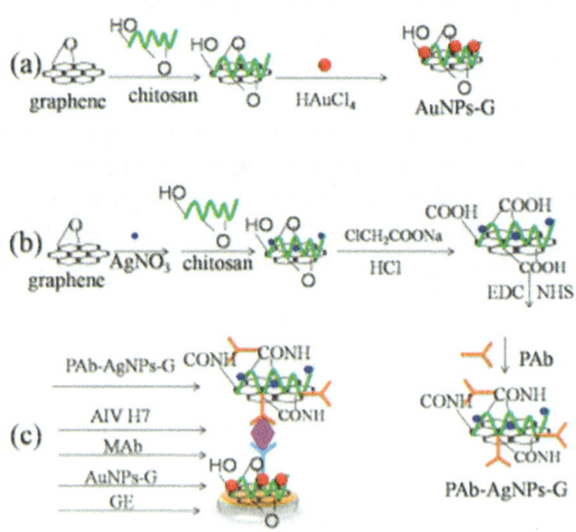

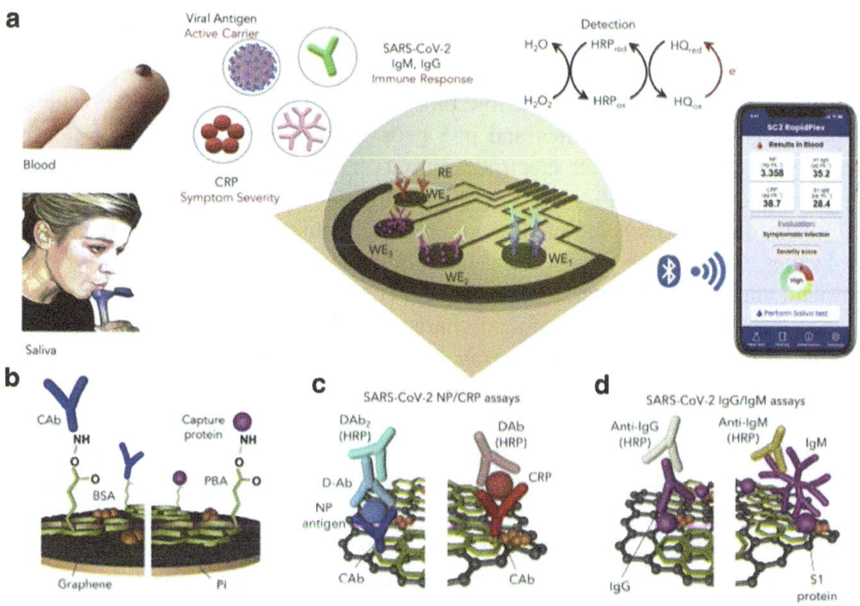

Fig. 4 **a** Graphene-based electrochemical platform to diagnose COVID-19 in serum and saliva using a portable and wireless system, **b** PBA interaction with both graphene electrode and biomolecules, **c** double-sandwich and sandwich configuration for NPr and CRP detection, respectively, and **d** indirect immunoassays for S1-IgG and S1-IgM detection [38]. Copyright 2021 (Reprinted with permission from Elsevier Ltd.)

Figure 4 illustrates the SARS-CoV-2 platform with four graphene working electrodes, which were fabricated via CO_2 laser engraving on a polyimide (PI) substrate. According to the authors, the graphene was chosen due to its superior properties, high charge mobility, and surface area, as well as high sensitivity and selectivity of sensing strategies involving both capture and detector receptors. In order to preserve the graphene structure, the surface was non-covalently functionalized with 1-pyrenebutyric acid (PBA) to anchor the required receptors, as shown in Fig. 4b. While the pyrene unit interacts with graphene by π-stacking and hydrophobic interactions, the carboxylic functional groups of PBA allow the covalently bound with amine groups of the specific antibodies or capture proteins. Bovine serum albumin (BSA) prevents non-specific adsorption of other biofluid molecules by reacting with unreacted sites on the graphene layer. The amperometric detections were performed based on the double-sandwich strategy for NPr, sandwich configuration for CRP (Fig. 4c), and indirect immunoassays for S1-IgG and S1-IgM (Fig. 4d). This is a promising approach for the COVID-19 diagnosis as graphene working electrodes can monitor four targets in one single and fast experiment (less than 1 min), portable and wireless system, in a non-invasive saliva test [38].

1.3 Other Biomolecules for Electrochemical Biorecognition

Avelino et al. [4] developed a flexible biosensor for label-free detection of human papillomavirus (HPV) families and p53 gene monitoring (biomarker of carcinogenicity) [4]. It used flexible electrodes based on polyethylene terephthalate (PET) strips coated with ITO. PPy-Au NPs films were electropolymerized on the flexible ITO electrode and then modified through the chemisorption of cysteamine (Cys) molecules, followed by the chemical immobilization of DNA-specific probes using covalent anchoring of glutaraldehyde (GA) [4].

According to the electrochemical characterization of the electrodes, PPy films provide a considerable number of nucleation sites for the formation of Au NPs with nanometric dimensions in a non-aggregation state. This surface allowed Cys molecules to function as a molecular spacer contributing to a higher conductivity response and lower interfacial resistance, increasing the sensitivity. The association of PPy and Au NPs allowed obtaining a platform with improved electrochemical properties and high surface area [4].

An alternative to horseradish peroxidase-based biosensors is using myoglobin, a heme-contained redox protein, to catalyze the hydrogen peroxide reduction. For example, an H_2O_2 electrochemical biosensor was developed using myoglobin-hybrid silica sol–gel coating on the surface of Ag NPs-doped CNTs/chitosan-modified electrodes [17]. Electron microscopy confirmed that the CNTs (diameter range 15–24 nm) prevented the Ag NPs (20–35 nm) agglomeration successfully. The high surface area of CNTs enhanced the electroactive area and, consequently, the background current in the presence of ferro/ferricyanide probe. As the biomolecule layer might function as a resistive component, the CNTs have an essential role in improving charge transfer. A synergetic effect between CNTs and Ag NPs during the detection of H_2O_2 in deoxygenated PBS resulted in improved electrocatalytic activity compared to the electrodes composed with only one of these components. The amperometric biosensor retained 78% of the initial activity after 50 days. This long-term stability was assigned to a proper microenvironment for myoglobin, provided by the film biocompatibility [17].

Sheng et al. [32] had explored the potential of nitrogen-doped CNTs combined with fullerene for the direct electrochemistry of hemoglobin. The fullerene is an excellent electron acceptor due to its unique dimensional and electronic structures. Thus, the combination of fullerene and ordered CNTs should facilitate the heterogeneous electron-transfer process. Additionally, the nitrogen-doped CNTs exhibit good electron conductivity, biocompatibility, and higher hydrophilicity than pristine CNTs, making them suitable to anchor biomolecules. When the hemoglobin was immobilized on fullerene/nitrogen-doped CNTs/chitosan, the apparent heterogeneous electron transfer rate constant suggested a fast electron transfer from the protein to the electrode because of the synergetic effect resulting from the carbon allotropes combination. As a result, the amperometric biosensor reached the plateaus within 5 s (response time) when applied on the electrocatalysis of H_2O_2 reduction [32].

2 Strategies for the Design of High-Performance Devices

2.1 *Machine Learning*

Artificial intelligence, especially machine learning (ML) algorithms, has been standing out as a fundamental tool for the design of high-performance electrochemical biosensors. However, despite all the advantages of electrochemical biosensors, they can also have some undesirable characteristics, such as their poor stability and short lifetime due to the instability of the biorecognition elements, besides their inevitably irregular signal to noise.

The capacity for processing extremely data set with ML algorithms enables a series of applications in the biosensor field, such as (i) categorization of the sensing signals into a series of categories based on the target analyte; (ii) correcting the biosensor response in the presence of interferents in real samples; (iii) reduction of noise, since the signal interference is widespread; (iv) object and pattern recognition; and (v) optimization of experimental conditions [7]. However, studies with ML to design organic semiconductor-based electrochemical biosensors are still very incipient.

The electrochemical biosensor output signal depends on the interaction of a sort of variables, such as electrical property (potential, current, impedance), characteristics of the electrode (support material, geometry, area), electrolyte composition (concentration, pH, solvent), external variables (temperature, pressure, time), and reaction variables (kinetic and thermodynamic parameters). Therefore, ML is a useful tool to predict the biosensor response through different experimental conditions, in addition, to optimizing the electrochemical biosensor response [10].

Recently, the literature reported the application of ML algorithms to obtain a classification model for a precise decision in the diagnosis of diseases such as endometriosis with a simple electrochemical immunosensor, allowing non-technical personnel to evaluate the result [25]. Also, ML algorithms were applied as a model to identify patterns of the analyte recognition (heavy metal salts, pesticides, and petrochemicals) in the response output, allowing to identify of these pollutants as well as their concentrations [1, 2].

In another work, an ML algorithm was used to predict the catechol concentration (a highly toxic compound for the environment and human body) using current as an input in voltammetric measurements using a glassy carbon electrode modified with PEDOT, graphene oxide nanosheets (GONs), and physically entrapped laccase (Lac) [21].

According to the literature, ML is helpful when the biomolecule activity decreases over time, requiring replacement intermittently. ML algorithms allow designing an intelligent electrochemical biosensor to predict the analyte concentration by considering the decrease of the electrochemical biosensor response over time [22].

The use of smart biosensor systems is growing especially in three areas: health and disease management [10], environmental pollution control [22], and agriculture and food science [1].

2.2 Screen-Printed Electrodes and Wearable Devices

The development of SPEs arrived to fill the need for small, portable, and disposable devices capable of accomplishing on-site analyses with high sensitivity, selectivity, a linear output, low power requirement, quick response, low cost, and multi-analyte detection [16]. This technology provides the necessary tool for developing PoC-based biosensors devices, besides the possibility of large-scale production. The use of SPE in PoC devices ranges on different applications, such as in the environmental (e.g., detection of pollutants, toxins, and microorganisms in water samples) [36] and in the health fields (e.g., monitoring and diagnosis of diseases, human metabolites, and drugs detection) [24].

The SPEs gather the three electrodes (reference, working, and counter) into a unique small system. They are fabricated by printing different viscous inks in successive steps onto a solid substrate, which can be a plastic film, a ceramic, a paper sheet, or a thin stretch film for instance. Usually, Ag/AgCl-based ink is used for the reference electrode; meanwhile, carbon ink is used for both counter and working electrodes. Carbon is used due to its excellent properties, such as low cost, wide potential window, low background currents, and chemical inertness [46].

The advantages of modifying the SPE with CPs and carbon-based materials are the improvements in conductivity, surface area, electron transfer rate, in addition to the suitable environment for the biomolecules. Biosensing electrodes for a series of target analytes can be developed by combining surface modification and SPE technology advantages.

Istamboulie et al. developed an enzymatic SPE for organophosphate pesticides detection [12]. A polyvinyl chloride (PVC) sheet was used as the screen-printed substrate, a glycerophthalic ink was used as an insulating layer, and Ag/AgCl was used as reference electrode. The carbon working electrode was modified with PEDOT:PSS to mediate the electron transfer reaction, and the AChE enzyme for the analyte detection, producing a biosensor with high conductivity and sensitivity for the detection of organophosphate pesticides.

The SPE permit, in addition to the electrode modifications, printing electrodes with different formats and shapes on varied materials. This technology has been employed to construct affordable wearable biosensors devices, which is very important in the healthcare field by allowing the continuous monitoring of several human metabolites (e.g., glucose, lactate, uric acid, and cholesterol), electrolytes (e.g., sodium and potassium), skin temperature, heart rate, blood pressure, oxygenation of the blood and brain activity, all in a non-invasive measurement, in real time, and continuous monitoring. These wearable devices can be bracelets, textile materials, and even tattoos over the epidermis [41]. Some of these devices require biocompatible materials, such as graphene and CNTs [6, 35, 40].

Gao et al. developed a flexible device that can be used as a wristband or a headband during physical activity for monitoring human metabolites, electrolytes, and skin temperature in sweat samples [9] using CNTs and PEDOT:PSS. In detail, it was used as a flexible PET substrate containing a printed sensing array (responsible for the analyte detection through the electrochemical signal generation) and a printed circuit board (responsible for signal transduction, conditioning, processing, and wireless transmission for a reading device, like a mobile phone or a computer).

For the sensor array fabrication, the PET was Cr/Au patterned with photolithography after the deposition of a parylene insulation layer to delimit the working electrode area, preventing electrical contact of the metal lines with skin and sweat. Ag/AgCl electrode was the reference electrode, and each sensing part was prepared according to its application: (i) glucose and lactate detection: Prussian blue was used as the electron transfer mediator and chitosan/CNTs/enzyme (GOx for the glucose detection and lactate oxidase for the lactate detection) for the biorecognition; (ii) sodium and potassium ions detection: a polyvinyl butyral (PVB) coated reference electrode was modified with PEDOT:PSS to act as an ion-to-electron transducer in this ion-selective electrode. The temperature sensor was fabricated using Cr/Au metal microwires. The developed sensor showed a great response, multi-analyte detection, specificity, and wearable technology.

The SPE is also a suitable technology to assemble tattoo devices. In this case, the biosensor may be printed on a tattoo paper and then transferred to the epidermis in order to measure target molecules in sweat or interstitial fluids. Jia et al. used this strategy to develop a temporary transfer tattoo-based electrochemical biosensor to detect lactate in human sweat [13]. The working electrode was modified with tetrathiafulvalene (TTF), CNTs, and the lactate oxidase enzyme. Finally, a biocompatible chitosan overlayer prevents the efflux of biochemicals in the reagent layer for the underlying epidermis. This non-invasive biosensor showed great response, specificity, and simplicity to operate.

Tracking circulating uric acid in biological fluids presents an enormous potential for reducing the risk of gout and hyperuricemia in clinical settings. It was recently reported a non-invasive wearable microfluidic-based electrochemical sensor capable of detecting uric acid in human sweat using PEDOT:PSS hydrogel in screen-printed carbon electrode (SPCE) with the conventional three-electrode system [45]. Copper was electrodeposited on the working electrode (Cu/SPCE) and then immersed in the PEDOT:PSS electrolyte (at a constant potential of 0.5 V during 450 s) for a complete and suitable layer thickness of the PEDOT:PSS hydrogel on the working electrode (PEDOT:PSS hydrogel/SPCE). For the fabrication of microfluidic devices, it was utilized two PDMS layers, one of them was laser engraved to form a microfluidic channel (sweat inlet and a reservoir) and the other to cover the channel surface to ensure the capillary effect. While the microfluidic device allows the real-time measurement, the PEDOT:PSS hydrogel provides conductivity, large electroactive surface area, electrolyte storing, enhanced flexibility, and distinctly enhanced catalytic activity for uric acid detection, as the oxidation peak current of uric acid increases and the oxidation peak potential reduces compared with the bare SPCE. This device was able to quantify uric acid in human sweat before and after consumption of a high-purine diet, showing its potential for the construction of high-performance wearable sensors.

The molecularly imprinted polymers (MIPs) can also be used in wearable biosensors that allow continuous monitoring of health status, reducing healthcare costs, time for diagnosis, and the need for centralized hospital-based care systems. For instance, a sensitive and selective biosensor for monitoring cortisol in human sweat was assembled by combining the MIPs and PEDOT:PSS [26]. Human sweat samples were collected during exercise. The sensor showed high chemical and physical stability at body temperature and resistance to physical deformation, presenting a log-linear response in the range of 0.01–10.0 μmol L^{-1}.

3 Conclusions

Organic semiconductors are suitable materials for the development of highly efficient biosensing devices. Their compositional and morphological versatilities allow the functionalization with different biomolecules for the detection of a variety of analytes with high specificity. Improvements in the performance may be achieved with a suitable electroanalytical apparatus based on the characteristics of the biorecognition element and the organic semiconductor. Progress toward miniaturized and wearable devices with improved biorecognition performance is expected with advances in tools and technologies such as machine learning, screen-printed electrodes, and microfluidics. Hopefully, non-invasive and not expensive devices will be available for society soon, bringing benefits to the future of diagnosis, environmental management, in addition to agriculture and food science.

References

1. Abu-Ali, H., Nabok, A., Smith, T. J.: Electrochemical inhibition bacterial sensor array for detection of water pollutants: artificial neural network (ANN) approach. Anal. Bioanal. Chem. **411**, 7659–7668 (2019). https://doi.org/10.1007/s00216-019-01853-8
2. Alonso, G. A., Istamboulie, G., Ramírez-García, A., Noguer, T., Marty, J. L, Muñoz, R.: Artificial neural network implementation in single low-cost chip for the detection of insecticides by modeling of screen-printed enzymatic sensors response. Comput. Electron. Agricult. **74**, 223–229 (2010). https://doi.org/10.1016/j.compag.2010.08.003
3. Anzai, J.-I.: Biosensors for the detection of OP nerve agents. In: Gupta, R.C. (eds) Handbook of Toxicology of Chemical Warfare Agents, 2nd ed., pp. 837–846. Academic Press (2015)
4. Avelino, K.Y.P.S., Oliveira, L.S., Lucena-Silva, N., Andrade, C.A.S., Oliveira, M.D.L.: Flexible sensor based on conducting polymer and gold nanoparticles for electrochemical screening of HPV families in cervical specimens. Talanta **226**, 122118 (2021). https://doi.org/10.1016/j.talanta.2021.122118
5. Aydemir, N., Malmström, J., Travas-Sejdic, J.: Conducting polymer based electrochemical biosensors. Phys. Chem. Chem. Phys. **18**, 8264–8277 (2016). https://doi.org/10.1039/C5CP06830D
6. Bianco, A., Kostarelos, K., Prato, M.: Making carbon nanotubes biocompatible and biodegradable. Chem. Commun. **47**, 10182–10188 (2011). https://doi.org/10.1039/C1CC13011K

7. Cui, F., Yue, Y., Zhang, Y., Zhang, Z., Zhou, H.S.: Advancing biosensors with machine learning. ACS Sens. **5**, 3346–3364 (2020). https://doi.org/10.1021/acssensors.0c01424
8. Cui, M., Song, Z., Wu, Y., Guo, B., Fan, X., Luo, X.: A highly sensitive biosensor for tumor maker alpha fetoprotein based on poly(ethylene glycol) doped conducting polymer PEDOT. Biosens. Bioelectron. **79**, 736–741 (2016). https://doi.org/10.1016/j.bios.2016.01.012
9. Gao, W., Emaminejad, S., Nyein, H.Y.Y., Challa, S., Chen, K., Peck, A., Fahad, H.M., Ota, H., Shiraki, H., Kiriya, D., Lien, D.-H., Brooks, G.A., Davis, R.W., Javey, A.: Fully integrated wearable sensor arrays for multiplexed in situ perspiration analysis. Nature **529**, 509–514 (2016). https://doi.org/10.1038/nature16521
10. Gonzalez-Navarro, F.F., Stilianova-Stoytcheva, M., Renteria-Gutierrez, L., Belanche-Muñoz, L.A., Flores-Rios, B.L. Ibarra-Esquer, J. E.: Glucose oxidase biosensor modeling and predictors optimization by machine learning methods. Sensors (Basel) **16**, 1483 (2016). https://doi.org/10.3390/s16111483
11. Huang, J., Xie, Z., Xie, Z., Luo, S., Xie, L., Huang, L., Fan, Q., Zhang, Y., Wang, S., Zeng, T.: Silver nanoparticles coated graphene electrochemical sensor for the ultrasensitive analysis of avian influenza virus H7. Anal. Chim. Acta **913**, 121–127 (2016). https://doi.org/10.1016/j.aca.2016.01.050
12. Istamboulie, G., Sikora, T., Jubete, E., Ochoteco, E., Marty, J.L., Noguera, T.: Screen-printed poly(3,4-ethylenedioxythiophene) (PEDOT): a new electrochemical mediator for acetylcholinesterase-based biosensors. Talanta **82**, 957–961 (2010). https://doi.org/10.1016/j.talanta.2010.05.070
13. Jia, W., Bandodkar, A.J., Valdés-Ramírez, G., Windmiller, J.R., Yang, Z., Ramírez, J., Chan, G., Wang, J.: Electrochemical tattoo biosensors for real-time noninvasive lactate monitoring in human perspiration. Anal. Chem. **85**, 6553–6560 (2013). https://doi.org/10.1021/ac401573r
14. Lalaoui, N., Elouarzaki, K., Goff, A.L., Holzinger, M., Cosnier, S.: Efficient direct oxygen reduction by laccases attached and oriented on pyrene-functionalized polypyrrole/carbon nanotube electrodes. Chem. Commun. **49**, 9281–9283 (2013). https://doi.org/10.1039/c3cc44994g
15. Li, L., Liang, B., Shi, J., Li, F., Mascini, M., Liu, A.: A selective and sensitive D-xylose electrochemical biosensor based on xylose dehydrogenase displayed on the surface of bacteria and multi-walled carbon nanotubes modified electrode. Biosens. Bioelectron. **33**, 100–105 (2012). https://doi.org/10.1016/j.bios.2011.12.027
16. Li, M., Li, Y.-T., Long, Y.-T.: Recent developments and applications of screen-printed electrodes in environmental assays—A review. Anal. Chim. Acta **634**, 31–44 (2012). https://doi.org/10.1016/j.aca.2012.05.018
17. Liu, C.-Y., Hu, J.-M.: Hydrogen peroxide biosensor based on the direct electrochemistry of myoglobin immobilized on silver nanoparticles doped carbon nanotubes film. Biosens. Bioelectron. **24**, 2149–2154 (2009). https://doi.org/10.1016/j.bios.2008.11.007
18. Liu, S., Ma, Y., Cui, M., Luo, X.: Enhanced electrochemical biosensing of alpha-fetoprotein based on three-dimensional macroporous conducting polymer polyaniline. Sens. Actuat. B Chem. **255**, 2568–2574 (2018). https://doi.org/10.1016/j.snb.2017.09.062
19. Loguercio, L.F., Thesing, A., Demingos, P., de Albuquerque, C.D.L., Rodrigues, R.S.B., Brolo, A.G., Santos, J.F.L.: Efficient acetylcholinesterase immobilization for improved electrochemical performance in polypyrrole nanocomposite-based biosensors for carbaryl pesticide. Sens. Actuat. B: Chem. **339**, 129875 (2021). https://doi.org/10.1016/j.snb.2021.129875
20. Majumdar, S., Mahanta, D.: Deposition of an ultra-thin polyaniline coating on a TiO_2 surface by vapor phase polymerization for electrochemical glucose sensing and photocatalytic degradation. RSC Adv. **10**, 17387–17395 (2020). https://doi.org/10.1039/D0RA01571G
21. Maleki, N., Kashanian, S., Maleki, E., Nazari, M.: A novel enzyme based biosensor for catechol detection in water samples using artificial neural network. Biochem. Eng. J. **128**, 1–11 (2017). https://doi.org/10.1016/j.bej.2017.09.005
22. Massah, J., Vakilian, K.A.: An intelligent portable biosensor for fast and accurate nitrate determination using cyclic voltammetry. Biosys. Eng. **177**, 49–58 (2019). https://doi.org/10.1016/j.biosystemseng.2018.09.007

23. Mazzei, F., Favero, G., Bollella, P., Tortolini, C., Mannina, L., Conti, M.E., Antiochia, R.: Recent trends in electrochemical nanobiosensors for environmental analysis. Int. J. Environ. Health **7**, 267–291 (2015). https://doi.org/10.1504/IJENVH.2015.073210
24. Mishra, R.K, Sempionatto, J.R., Li, Z., Brown, C., Galdino, N.M. Shah, R., Liu, S., Hubble, L.; J., Bagot, K., Tapert, S., Wang, J.: Simultaneous detection of salivary Δ9-tetrahydrocannabinol and alcohol using a wearable electrochemical ring sensor. Talanta **211**, 120757 (2020). https://doi.org/10.1016/j.talanta.2020.120757
25. Pal, A., Biswas, S., Kare, S.P.O., Biswas, P., Jana, S.K., Das, S., Chaudhury, K.: Development of an impedimetric immunosensor for machine learning-based detection of endometriosis: a proof of concept. Sens. Actuat. B Chem. **346**, 134060–134073 (2021). https://doi.org/10.1016/j.snb.2021.130460
26. Parlak, O., Keene, S.T., Marais, A., Curto, V.F., Salleo, A.: Molecularly selective nanoporous membrane-based wearable organic electrochemical device for noninvasive cortisol sensing. Sci. Adv. **4**, eaar2904 (2018). https://doi.org/10.1126/sciadv.aar2904
27. Pundir, C.S., Chauhan, N.: Acetylcholinesterase inhibition-based biosensors for pesticide determination: a review. Anal. Biochem. **429**, 19–31 (2012). https://doi.org/10.1016/j.ab.2012.06.025
28. Putzbach, W., Ronkainen, N.J.: Immobilization techniques in the fabrication of nanomaterial-based electrochemical biosensors: a review. Sensors. **13**, 4811–4840 (2013). https://doi.org/10.3390/s130404811
29. Ramanavicius, S., Ramanavicius, A.: Conducting polymers in the design of biosensors and biofuel cells. Polymers **13**, 49 (2021). https://doi.org/10.3390/polym13010049
30. Ronkainen, N.J., Halsall, H.B., Heineman, W.R.: Electrochemical biosensors. Chem. Soc. Rev. **39**, 1747–1763 (2010). https://doi.org/10.1039/b714449k
31. Saleem, W., Salinas, C., Watkins, B., Garvey, G., Sharma, A.C., Ghosh, R.: Antibody functionalized graphene biosensor for label-free electrochemical immunosensing of fibrinogen, an indicator of trauma induced coagulopathy. Biosens. Bioelectron. **86**, 522–529 (2016). https://doi.org/10.1016/j.bios.2016.07.009
32. Sheng, Q., Liu, R., Zheng, J.: Fullerene-nitrogen doped carbon nanotubes for the direct electrochemistry of hemoglobin and its application in biosensing. Bioelectrochemistry **94**, 39–46 (2013). https://doi.org/10.1016/j.bioelechem.2013.05.003
33. da Silva, M.K.L., Vanzela, H.C., Defavari, L.M., Cesarino, I.: Determination of carbamate pesticide in food using a biosensor based on reduced graphene oxide and acetylcholinesterase enzyme. Sens. Actuat. B Chem. **227**, 555–561 (2018). https://doi.org/10.1016/j.snb.2018.09.051
34. Soganci, T., Soyleyici, H.C., Demirkol, D.O., Ak, M., Timur, S.: Use of super-structural conducting polymer as functional immobilization matrix in biosensor design. J. Electrochem. Soc. **165**, B22–B26 (2018). https://doi.org/10.1149/2.0561802jes
35. Syama, S., Mohanan, P.V.: Safety and biocompatibility of graphene: a new generation nanomaterial for biomedical application. Int. J. Biol. Macromol. **86**, 546–555 (2016). https://doi.org/10.1016/j.ijbiomac.2016.01.116
36. Taleat, Z., Khoshroo, A., Mazloum-Ardakani, M.: Screen-printed electrodes for biosensing: a review (2008–2013). Microchim. Acta **181**, 865–891 (2014). https://doi.org/10.1007/s00604-014-1181-1
37. Tang, W., Li, L., Zeng, X.: A glucose biosensor based on the synergistic action of nanometer-sized TiO_2 and polyaniline. Talanta **131**, 417–423 (2015). https://doi.org/10.1016/j.talanta.2014.08.019
38. Torrente-Rodríguez, R.M., Lukas, H., Tu, J., Min, J., Yang, Y., Xu, C., Rossiter, H.B., Gao, W.: SARS-CoV-2 RapidPlex: a graphene-based multiplexed telemedicine platform for rapid and low-cost COVID-19 diagnosis and monitoring. Matterial **3**, 1981–1998 (2020). https://doi.org/10.1016/j.matt.2020.09.027
39. Wang, J.: Electrochemical glucose biosensors. Chem. Rev. **108**, 814–825 (2008). https://doi.org/10.1021/cr068123a

40. Wang, K., Ruan, J., Song, H., Zhang, J., Wo, Y., Guo, S., Cui, D.: Biocompatibility of graphene oxide. Nanoscale Res. Lett. **6**, 8 (2011). https://doi.org/10.1007/s11671-010-9751-6
41. Windmiller, J.R., Wang, J.: Wearable electrochemical sensors and biosensors: a review. Electroanalysis **25**, 29–46 (2013). https://doi.org/10.1002/elan.201200349
42. Wong, C.M., Wong, K.H., Chen, X.D.: Glucose oxidase: natural occurrence, function, properties and industrial applications. Appl. Microbiol. Biotechnol. **78**, 927–938 (2008). https://doi.org/10.1007/s00253-008-1407-4
43. Xia, H., Zeng, J.: Rational surface modification of carbon nanomaterials for improved direct electron transfer-type bioelectrocatalysis of redox enzymes. Catalystsis **10**, 1447 (2020). https://doi.org/10.3390/catal10121447
44. Xie, Q., Weng, X., Lu, L., Lin, Z., Xu, X., Fu, C.: A sensitive fluorescent sensor for quantification of alpha-fetoprotein based on immunosorbent assay and click chemistry. Biosens. Bioelectron. **77**, 46–50 (2016). https://doi.org/10.1016/j.bios.2015.09.015
45. Xu, Z., Song, J., Liu, B., Lv, S., Gao, F., Luo, X., Wang, P.: A conducting polymer PEDOT:PSS hydrogel based wearable sensor for accurate uric acid detection in human sweat. Sens. Actuat. Chem. **348**, 130674 (2021). https://doi.org/10.1016/j.snb.2021.130674
46. Yamanaka, K., Vestergaard, M.C., Tamiya, E.: Printable electrochemical biosensors: a focus on screen-printed electrodes and their application. Sensors **16**, 1761 (2016). https://doi.org/10.3390/s16101761
47. Yang, S., Liu, J., Zheng, H., Zhong, J., Zhou, J.: Simulated revelation of the adsorption behaviours of acetylcholinesterase on charged self-assembled monolayers. Nanoscale **12**, 3701–3714 (2020). https://doi.org/10.1039/c9nr10123c

Graphene-Based Bioelectronics

Isabela Alteia Mattioli and Frank N. Crespilho

1 Introduction

In this chapter, we will discuss about graphene-based bioelectronics and its impacts on biosensing and analytical detection research. For this, aspects concerning graphene obtention and its intrinsic properties will be glimpsed. We will move then towards graphene applications in bioelectronics, giving firstly an introduction on what is bioelectronics and its impacts on emerging technologies. Sequentially, some graphene electronic properties and their impact on the performance of a bioelectronic device will be discussed. Then, graphene field-effect transistors for bioelectronics purposes will be presented and detailed. In this context, functionalization strategies for this sort of device will be in-depth debated. Finally, modern applications of field-effect graphene devices for bioelectronics will be presented and discussed.

2 Graphene: Overview and Obtention

Graphene is one of the various carbon allotropes existent in nature. Consisting of a two-dimensional (2D) carbon atoms network organized in a honeycomb lattice, graphene has been first isolated in 2004 by Geim and Novoselov [1]. Since then, this 2D crystalline material with one-atom width has been extensively studied in research community aiming applications in (bio)electronics, nanomedicine, diagnosis, electrochemistry and portable energy storage [2]. The remarkable attention

I. A. Mattioli · F. N. Crespilho (✉)
Sao Carlos Institute of Chemistry, University of Sao Paulo, Av. Trabalhador Sao-Carlense, 400, Sao Carlos, SP 13560-970, Brazil
e-mail: frankcrespilho@iqsc.usp.br

I. A. Mattioli
e-mail: isabela.mattioli@usp.br

F. N. Crespilho (ed.), *Advances in Bioelectrochemistry Volume 4*,
https://doi.org/10.1007/978-3-030-99662-8_8

that graphene has attracted is based on its unique thermal, mechanical, optical and electronic properties [2], as high optical transparency (up to 97.7%) [3], mechanical strength (42 N m^{-1}) [4], improved thermal conductivity (5000 W m^{-1} K^{-1}) [3], low electrical resistance (<10 kΩ sq^{-1}) [5] and high charge carrier mobility (200,000 cm^2 V^{-1} m^{-1}) [6].

Although mechanical exfoliation (i.e. "Scotch-Tape Method") was the first developed route for obtaining graphene from graphite with simplicity and low-cost features [1, 2], nowadays, there are other sophisticated and highly controlled methods for graphene synthesis. In particular, these techniques are useful for obtaining low-defective graphene structures specially for optical and electronic applications, as we will see in the further sections. In this context, chemical vapor deposition (CVD) is the most common route for graphene obtention [7]. It consists of graphene growth on a metallic substrate (mainly Cu) after thermal decomposition of a carbon source molecule. For this, ethylene gas is frequently used and its adsorption onto the metallic substrate under high temperature (above 1000 °C) leads to the growth of graphene single crystalline domains, forming polycrystalline graphene adsorbed on Cu foil [2, 7].

Another interesting methodology is graphene growth on SiC substrate, producing epitaxial graphene. Briefly, SiC substrates undergo high temperatures (>1400 °C) at high vacuum and carbon atoms are rearranged at the surface as silicon atoms evaporate. At the end of this process, graphitic layers are formed. Although the production of graphene layers directly on Si substrate (the most applied substrate for graphene electronics and bioelectronics) is a main advantage, this process is considered highly costly [8]. Recently, laser-induced methodologies have been explored in literature for the production of graphene with high yield rates [4]. A carbon-based precursor material rich in sp^3-hybridized atoms is irradiated with infrared laser pulses, resulting in the formation of sp^2 hybridized carbon atoms that are rearranged in a honeycomb lattice, leading to graphene [9]. To become this process more cost-effective, literature has reported studies on the application of commercial polymers as carbon-based precursor material [10].

3 Graphene for Bioelectronics

3.1 Introduction to Bioelectronics

Bioelectronics aims sensing and actuating by two-dimensional thin films and entities. These devices allow the real-time monitoring of physiological processes as well as medical diagnosis without need for further body interventions. For these purposes, bioelectronics is of great interest in healthcare, biologic research and nanomedicine [4].

However, there are some key features that 2D materials should present in order to be considered a suitable material for bioelectronics, as chemical and physical stability, high specific surface area, planal material topography, conductivity, electrochemical features and biocompatibility [4]. Each of this feature has a direct impact on the bioelectronic device performance. Material topography can influence on the biological specie adhesion onto the device, and therefore, on the magnitude of transduced signal [11]. Biocompatibility is of great importance for properly interfacing with biological environments as cells and tissues [4]. Chemical and physical stabilities are important parameters for the development of robust bioelectronic devices able to be applied in diverse conditions in biological environments without production on undesired secondary species. Specific surface area, by its turn, is desirable for improving the analytical performance of the device. Literature usually correlates devices based on materials with higher specific surface area with the ability to reach lower limits of detections (LOD) [4, 12]. This can be due to the increase in electrochemically active area of the device (if it is operated by electrochemical principles), and/or increase in surface area available for adsorbing the biological target analyte.

3.2 *Graphene Properties for Bioelectronics Purposes*

There are a wide range of materials that can fit the above-mentioned requirements for a successful device design. Semiconductors [13], metals [14] and graphene-based materials [15, 16] are the most applied materials in literature bioelectronic device reports [4]. Amongst them, graphene is one of the most explored nanomaterials for bioelectronics purposes due to its unique physico-chemical properties. For bioelectronics, pristine graphene electronic and electrical properties are of greater interest, as these types of devices concern the synergy between electronics and biology [17]. Thus, charge carrier mobility, basal plane conductivity, low intrinsic resistivity, ambipolar characteristics and high charge carrier density at its Charge Neutrality Point are a few to mention [2, 15, 18]. Another interesting aspect of graphene is its "band-gap zero semiconductor features" at 0 K, contrasting to every other material that may be used for electronics or bioelectronics applications [2].

All these properties arise due to graphene's unique energy band structure and can be correlated to its linear energy dispersion and linear Density of States (DOS) distribution [2, 15]. For a better understanding of this energy band structure, it is important to consider some symmetry properties that can be analyzed in Brillouin zones at reciprocal space (Fig. 1).

At graphene's first Brillouin zone in reciprocal space, graphene presents several high-symmetry points denominated K and K′. These points are known as Charge Neutrality Point (CNP) or Dirac Point, where a null DOS is observed (Fig. 2). This point is also referred as Fermi level in literature and lies at the intersection between two cone-like band structures related to filled valence and empty conduction bands [15, 20]. At this point, it is reported that electrons move with Fermi velocity. Moreover, an equal density of both electrons and holes is found, with high charge carrier density (around 10^{-12} cm^{-2}) without the application of external gate voltages or other doping mechanisms (i.e. graphene with pristine characteristics) [15].

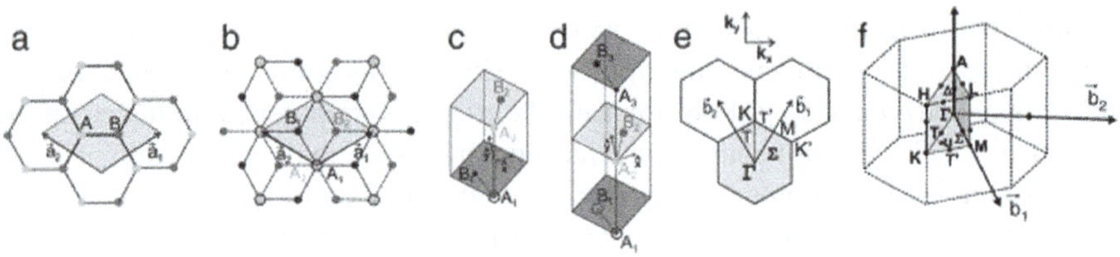

Fig. 1 **a** Unit cell of graphene monolayer (top view); **b** bilayer graphene real space image, with upper (gray) and lower (black) atoms; **c** unit cell for bilayer graphene; **d** unit cell for trilayer graphene; **e** reciprocal space unit cell in first Brillouin zone for monolayer graphene showing high-symmetry points and **f** Brillouin reciprocal space unit cell presenting high symmetry points. Reprinted from reference [19] with permission from Elsevier

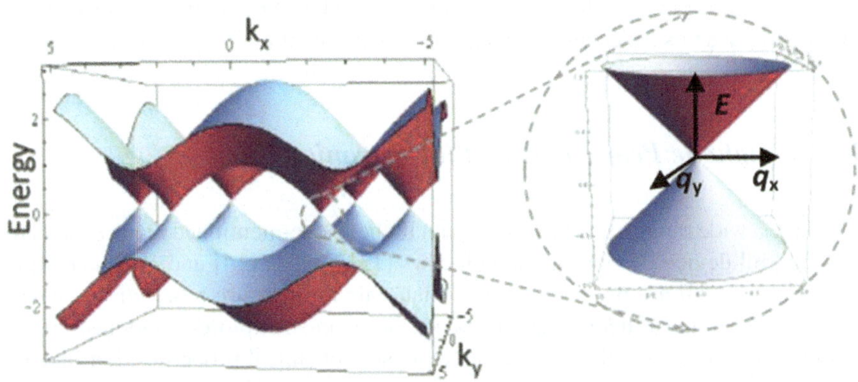

Fig. 2 Graphene tight binding-based band structure (left) and low energy band structure near charge neutrality point band structure. Reprinted from reference [20] with permission from IOP Publishing

It is important to remember that gate voltages are external electrical field polarizations that induce conductivity changes at graphene basal plane [18]. Away from this point, it is observed a linear dependence on the charge carrier density with an external gate voltage that can be applied for modulation of electric field and conductivity improvements [15]. The insertion of charge carriers (n) can be described by $n = V_g \cdot \alpha$, where α is charge injection rate, and V_g is gate voltage [15]. Based on this, the graphene ambipolar behavior can be observed, once electrons and holes can be continuously changed as predominant charge carriers by the application of a positive or negative gate voltage. Insights on how carrier density is being modulated by the application of a gate voltage can be given by analyzing the Holzer coefficient, defined by $R_H = 1/(e \cdot n)$. Graphene resistivity (ρ) is also affected by the amount of charge carrier inserted at graphene electronic structure, as well as its mobility through basal plane (μ), as demonstrated by the relation $\rho = 1/en\mu$ [15].

Charge carrier mobility is another important graphene intrinsic property derived from its energy band structure and honeycomb lattice. Literature has reported that graphene can reach 20,000 $cm^2\ V^{-1}\ s^{-1}$ when supported on Si at 280 K [21], and up to 200,000 $cm^2 V^{-1} s^{-1}$ when suspended around 150 nm above a Si/SiO_2 gate electrode, with a charge carrier density of 5×10^{11} cm^{-2} [6]. Moreover, literature has reported that chemically or physically doped graphene devices still present exceptional charge carrier mobilities, evidencing the occurrence of ballistic transport [22]. Charge carrier mobility is of particular interest for high-speed graphene electronic components and bioelectronics in which working principles rely on the current passage through graphene basal plane. The fact that it is weakly dependent on temperature [22] is another advantage for producing devices and components for operation out of room temperature. However, the existence of grain boundaries in graphene is reported to scatter charge carriers and reduce their mobility in its surroundings up to ~5.9% [23]. This is a common disadvantage of large-area graphene devices (i.e. millimetric-sized), and therefore miniaturization of these devices down to micrometer or nanometer scales has already been widely studied [24–26].

3.3 Graphene Field-Effect Transistors

The above-discussed properties are of great interest for electronic graphene devices, especially GFETs electronic devices with three terminals: source, drain and gate. GFET working principles rely on the passage of current between drain and source terminals through graphene's basal plane, under an applied voltage namely drain-to-source voltage (V_{ds}), while an independent gate voltage (V_g) is simultaneously applied for electrical field modulation [18, 20]. Therefore, for these devices to work properly, the employed graphene properties mentioned in the section above must be explored, mainly high charge carrier mobility, high basal plane conductivity, low intrinsic resistivity and high charge carrier density. In this sense, GFETs have received much attention from research and innovation community for high-frequency electronics applications owing to its, low power operation, satisfactory scalability and high current carrying capability [20]. Other GFET features are of great interest for graphene-based bioelectronics, as high sensitivity, returning high current or voltage changes in response to trace levels chemical and physical stability and ability to reach low LODs are few to mention [16]. Another important aspect inherent to pristine monolayer graphene concerns is its transconductance (g_m), mainly explored in GFETs. It consists of the variation rate of drain-source current (I_{ds}) in relation to V_g, as described by the expression $g_m = dI_{ds}/dV_g$ [27] and gives information on the device amplifying ability [28]. In other words, a GFET with optimized transconductance would present high sensitivity, as a remarkable voltage gain or an interesting percent change of GFET current, defined by dI_{ds}/I_{ds} [16, 27].

Another GFET feature of great interest for highly sensitive bioanalyte detection is the signal-to-noise ratio (SNR). It is an indicative of the GFET reliability and reflects the ratio between the transduced net signal and the background signal [16]. Particularly, the background signal depends on some matrix signals, as the Debye length and operational GFET parameters, as frequency [16, 29]. Moreover, it is also reported that biosensing near CNP is performed with reduced background electrical noise, and this strategy may be one of the more advantageous for biosensing of compounds at trace levels [29].

On the other hand, there are some GFET properties that are not as advantageous as desirable for bioelectronics. For example, due to the high interfacial sensitivity to chemical environment alterations of these devices (i.e. adsorption of any biomolecule onto graphene, or pH changes), GFETs often present low specificity. To overcome this challenge, graphene functionalization is frequently required.

3.4 Graphene Functionalization for Bioelectronics Purposes

Graphene functionalization is a widely applied strategy for improving some graphene properties and produces a better-fitted graphene device for the desired application. In biosensing, sensitivity, lower LOD, specificity and detection performance are the most explored features through functionalization.

Functionalization can be made through covalent bonding our non-covalent interactions. Covalent methods for graphene modification involve sp^2 lattice rupture and a variety of functional groups insertion, depending on the modifier precursor [30]. These groups are considered defects on graphene structure and can be monitored by Raman Spectroscopy, by evaluating an increase in D band intensity in relation to G band one (ratio I_D/I_G). Thus, an increase on I_D/I_G ratio is a rough indicative of successful functionalization [30, 31].

There are different covalent functionalization methods. One of the most employed in graphene research and applications regards the production of graphene derivatives, as graphene oxide (GO) and reduced graphene oxide (rGO). In both cases, the insertion of oxygenated functional groups in both graphene edge and basal plane is performed. In rGO, particularly, the reduction of a fraction of these groups is made through electrochemistry, chemical oxidation or thermal oxidation [32]. Despite the basal plane conductivity loss due to covalent functionalization, literature contain some reports on the use of GO and rGO for field-effect-based devices [33–35]. These materials are also frequently employed for graphene electrochemical devices, as an alternative to the electrochemical sluggish electron transfer kinetics presented by pristine monolayer graphene [36]. It is known that GO and rGO present enhanced electrochemical activity due to the presence of oxygenated functional groups capable of realizing off-plane electron transfer and taking part into electrochemical reactions [32]. Another sort of graphene functionalization widely explored concerns the use of diazonium derivatives aiming the insertion of different desirable functional groups. This strategy is of great interest for biosensing purposes due to the possibility of attaching a variety of suitable functional groups capable of anchoring large biomolecules, as proteins [37, 38]. Despite this advantage, this sort of functionalization was reported to decrease significantly charge carrier mobility and saturation currents in diazonium derivative-functionalized GFETs based on monolayer graphene sheets [39]. As an alternative, the use of GFETs relying on bilayer pristine graphene appeared to contour these problems and increase the robustness of this bioelectronic device [39]. Still regarding covalent functionalizations, there are reports in literature describing the possibility of directioning of covalent functionalization in graphene devices (Fig. 3) [40–43]. This is of great usefulness for producing highly sensitive, miniaturized and single selective bioelectronics assays [42]. Generally, graphene edges are found to have a superior reactivity in comparison to basal plane, and functional group insertion occurs preferentially in these regions [42].

Due to sp^2 lattice disruption, covalent functionalization is frequently avoided for electrical-based graphene devices, as GFETs. This is because the basal plane sp^2 lattice rupture due to functional groups insertion affects graphene electrical properties, by decreasing basal plane conductivity and charge carrier mobility, as well as the opening of a band-gap [18, 30]. As a consequence, current passage through drain and source terminals is strongly affected [44]. To overcome these challenges, non-covalent functionalization aroused as suitable modification alternatives for GFETs and other electrical-based graphene devices. It is based on the establishment of van der Waals interactions between graphene π electronic structure and the modifier, leading to a π–π stacking [2, 18, 45]. As an example, graphene functionalization with ferrocene through van der Waals interactions was largely studied in literature [46–48] and was reported to lead to a promising heterojunction for genetic material detections

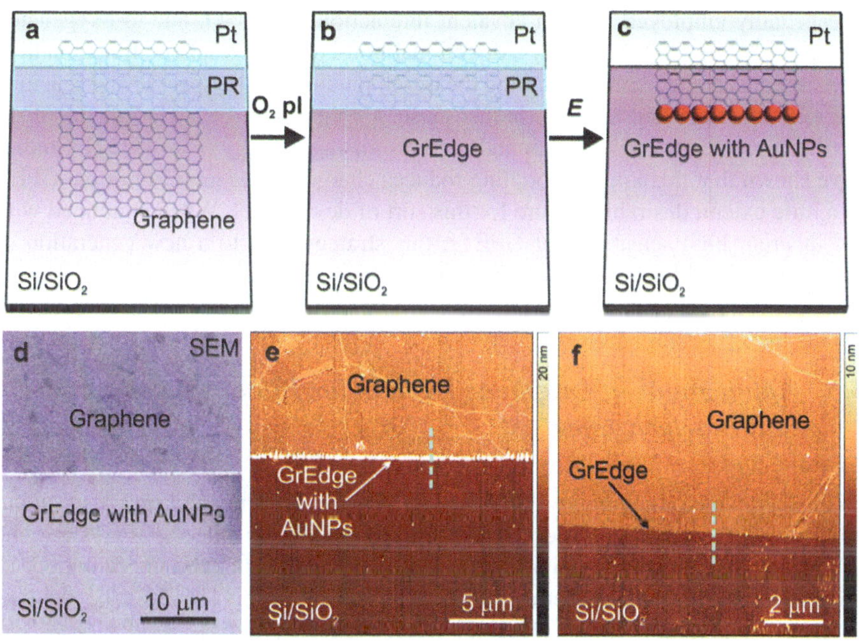

Fig. 3 **a** Photoresist patterning of graphene monolayer; **b** etching of uncovered regions with O_2 plasma; **c** AuNP electrodeposition at graphene edges and photoresist dissolution; **d** scanning electron microscope images of AuNP-modified graphene edge; **e** atomic force microscopy images of AuNP functionalized graphene edge in comparison to **f** bare graphene edge region. Reprinted from reference [42] with permission from Royal Society of Chemistry

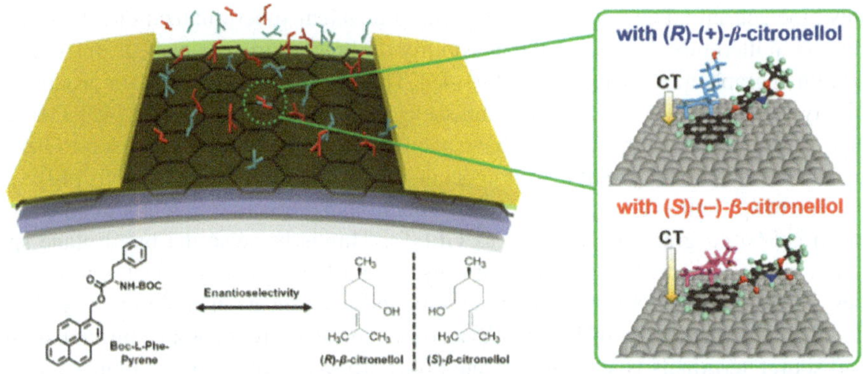

Fig. 4 Schematic representation of non-covalent functionalization of GFET devices with a new synthesized compound (Boc-L-Phe-Pyrene) for the enantioseletive detection of (*R*)-(+)-β-Citronellol and (*S*)-(-)-β-Citronellol. Reprinted from reference [52] with permission from American Chemical Society

[18, 45]. Non-covalent functionalization with polymers for biomolecules anchoring is also widely explored, as PBASE [49] and poly-vinyl-pyrrolidone [50]. APTES is especially employed for non-covalent interactions with GO, due to oxygenated groups coupling [51]. Nowadays, directional non-covalent graphene functionalizations are also being explored, aiming the development of more specific and sensitive graphene bioelectronic devices. In this sense, graphene non-covalent functionalization with enantioselective compounds is a promising strategy (Fig. 4) [52]. Studies have shown that this kind of coupling reduces charge carrier mobility of the GFET in a little extent, desirable feature for this sort of device [52]. When combined with the inherent high sensitivity of GFETs, this strategy lead to a new generation of miniaturized highly specific and sensitive devices.

3.5 Applications of Graphene Devices for Biosensing and Bioelectronics

So far, we have talked about graphene electronic properties and how they affect and influence graphene bioelectronics, especially GFETs. Indeed, GFETs are one of the most promising graphene bioelectronic devices for biosensing a variety of biomolecules, as genetic material [29, 49, 53–57], biomarkers [35, 58–60], enzymes [34, 61, 62] and, nowadays, COVID-19-related species [63–65].

There is a large amount of proteins detected by GFETs in literature, as immunoglobulins, biomarkers and enzymes are within this class. As an example, back in 2010, Ohno and collaborators reported a non-modified GFET capable of recognizing Bovine Serum Albumine (BSA) through adsorption mechanisms at graphene's surface. These studies made with BSA in nmol L^{-1} concentration order, proved that

GFETs could be promising candidates for biosensing purposes [66]. As bioelectronics research evolved, more complex GFET-based systems began to be explored. For increasing sensitivity, strategies of immobilization of specific probes began to be explored. As an example, protein-antibody affinity interactions were studied in a gold-coated GFET, with great affinity for thiol-based cross-linker molecules [67]. Likewise, aptamer-protein binding kinetics on a GFET platform was also studied [68]. This strategy concerns a two-step cross-linking process [69] and has been frequently employed in literature [68–72], as it confers high specificity. Besides this, there are other increasing specificity strategies of great usefulness for the detection of other proteinaceous species by GFETs, as immunoglobulins and biomarkers. For example, carcinoembryonic antigens (CEA) were detected by a GFET modified with EDC, NHS and nano-denaturated BSA, reaching CEA concentrations of 337.58 fg mL^{-1} [58]. The nano-denaturated BSA film deposited onto graphene modified with EDC and NHS acted as a cross-linker to immobilized antibodies for specific interaction with CEA (Fig. 5) [58]. Another example of successful biomarker detection through GFETs concerns the use of a rGO GFET modified with PBASE for the detection of Alzheimer's disease biomarkers [35]. In this study, pico to nanomolar concentrations could be analyzed in a multiplexed platform based on CNP shifts due to biomarker adsorption [35].

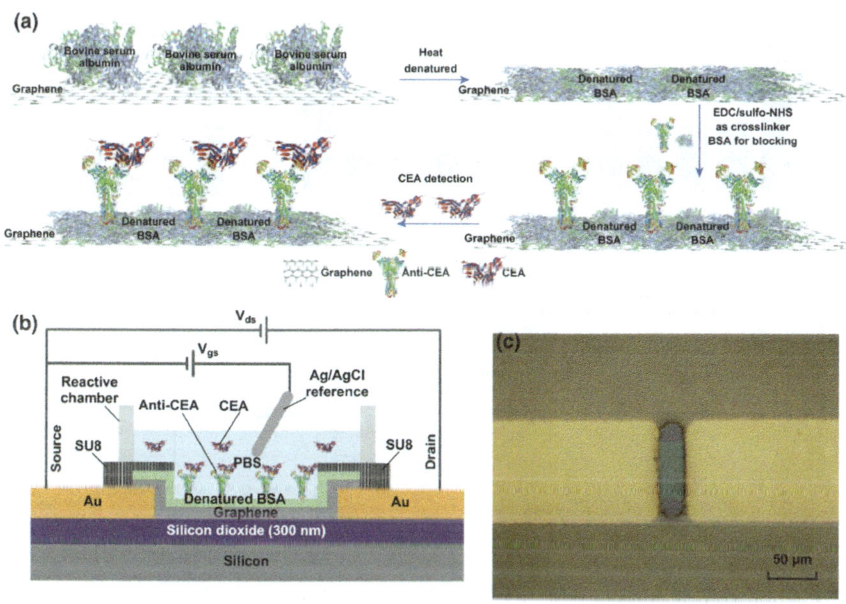

Fig. 5 **a** Steps involved in the GFET modification with anti-CEA antibodies and nanodenaturated BSA; **b** schematic illustration of the final device for CEA detections; **c** optical image of graphene channel of this device. Reprinted from reference [58] with permission from Springer Nature, under CC BY license

Graphene bioelectronics for genetic material is also widely explored in literature. The most recent studies report femto to zeptomolar detections of DNA through different strategies. As an example, DNA zeptomolar detections were performed in a crumpled GFET after processing this genetic material in a LAMP assay for signal amplification (Fig. 6) [73]. Despite of the remarkable achieved analytical parameters, the proposed assay does not present much simplicity. On the other hand, DNA detections could also be performed by exploring DNA and PNA hybridization strategies, however, with a less prominent LOD of 100 fmol L^{-1} [55]. For this, a rGO-based GFET was employed, instead of pristine GFET. The identification of individual DNA nucleobases was also explored by GFETs. This could be made due

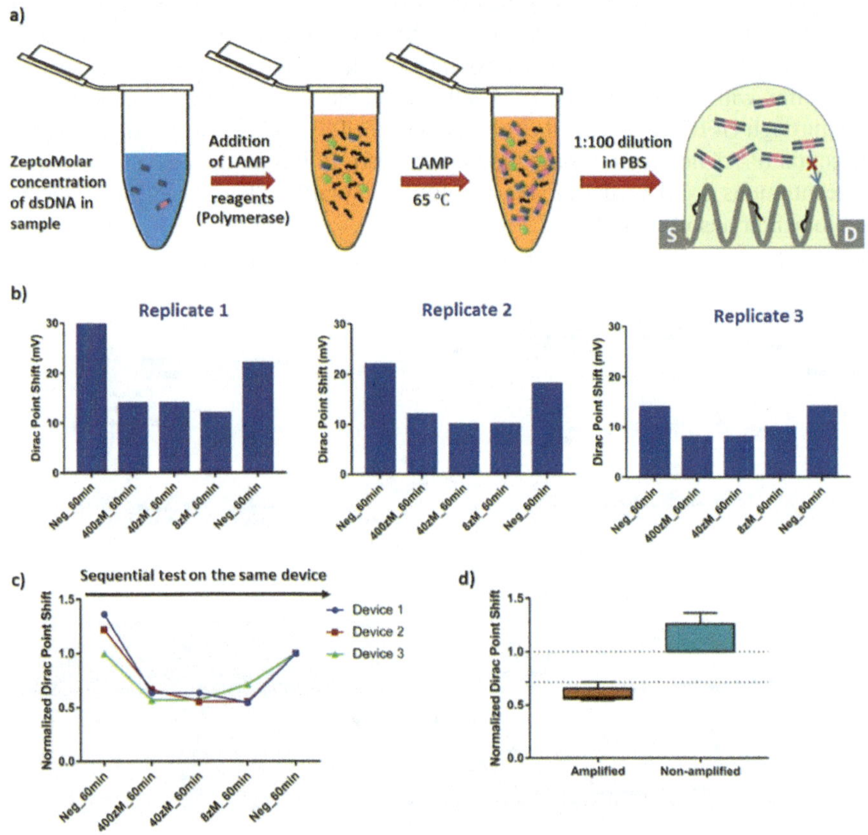

Fig. 6 Zepto-molar LAMP reactions prior to detections using crumpled GFET devices; b) Negative control (primer) and amplified ds DNA samples absolute CNP voltages shifts; c) Studies on the performance of sequential tests using the same device by evaluation of returned CNP voltages with negative control (primer) and amplified ds-DNA samples; d) Normalized CNP shifts for amplified and non-amplified ds-DNA samples detections, $n = 3$. Reprinted from reference [73] with permission from WILEY-VCH Verlag GmbH & Co. KGaA

to different interfacial dipole fields that led to individual conductance signatures for each nucleobase when directly deposited on pristine graphene channel [57].

Currently, the COVID-19 pandemic outbreak revealed the necessity of exploring bioelectronics as rapid, sensitive and efficient diagnosis tools. In this sense, GFETs started to be employed in literature especially for viral SARS-CoV-2 detections, aligning great sensitivity with rapidness of analyses. Moreover, strategies for specificity have been adopted, to contour one of the most recurrent GFETs problematics. As an example, Seo and collaborators developed the first GFET for SARS-CoV-2 viral detections. For this, specific antibodies for S protein were immobilized onto graphene by PBASE crosslinking, and fg mL^{-1} levels of S protein concentration could be successfully detected [65]. Viral RNA detections in clinical samples were also explored by GFETs. In this case, gold nanoparticles and phosphorodiamidate morpholino oligos (PMO) probes were deposited onto the GFET surface [63]. In this case, PMO was proved to be a more suitable probe for viral RNA than ssDNA, with negligible background signal [63]. 2D transition metal carbides, known as MXenes, were also explored together with GFETs towards COVID-19 diagnosis in the study developed by Li and collaborators [64]. In this work, the use of a MXene could confer superior chemical sensitivity and lower sheet resistance, allowing to perform highly sensitive detections with an easily measurable signal. Using this MXene GFET, SARS-CoV-2 and H1N1 antigens could be differentiated and SARS-CoV-2 detections in fg mL^{-1} concentration levels could be performed (Fig. 7) [64].

3.6 Graphene Electrical-Electrochemical Vertical Devices

So far, the advantages and applicabilities of GFETs have been pointed out in the previous sections. However, GFETs have some challenges to be overcome. One of the most underdiscussed examples concerns the use of an $Ag/AgCl_{sat}$ electrode as gate electrode for potential reading in I_{ds} versus V_g curves. In this sort of application, the $Ag/AgCl_{sat}$ electrode is polarized for V_g sweepings [18, 20]. Nonetheless, its equilibrium system established between Ag^+ and AgCl into the electrode body makes this electrode as an ideally non-polarizable electrode. Its potential contribution to the electrochemical systems in which it is used is approximately 0.197 V (vs. SHE) [74]. Therefore, the use of $Ag/AgCl_{sat}$ reference electrode as gate electrode may cause instabilities in the electrode internal equilibrium, which could lead to imprecisions on potential readings during V_g sweeping as well as shifts in its equilibrium potential [74]. To avoid these problems, recently, Mattioli, Crespilho and collaborators studied the application of a monolayer graphene onto Si/SiO_2 device submitted to hybrid electrical-electrochemical working principles, generating the Electrical-Electrochemical Vertical Devices (EEVDs, Fig. 8) [18]. These devices appeared to be promising alternative assays for biosensing purposes with miniaturized features and possibility of reaching unprecedently low LODs for genetic material detections [18]. Due to this, intellectual protection of this technology is on process.

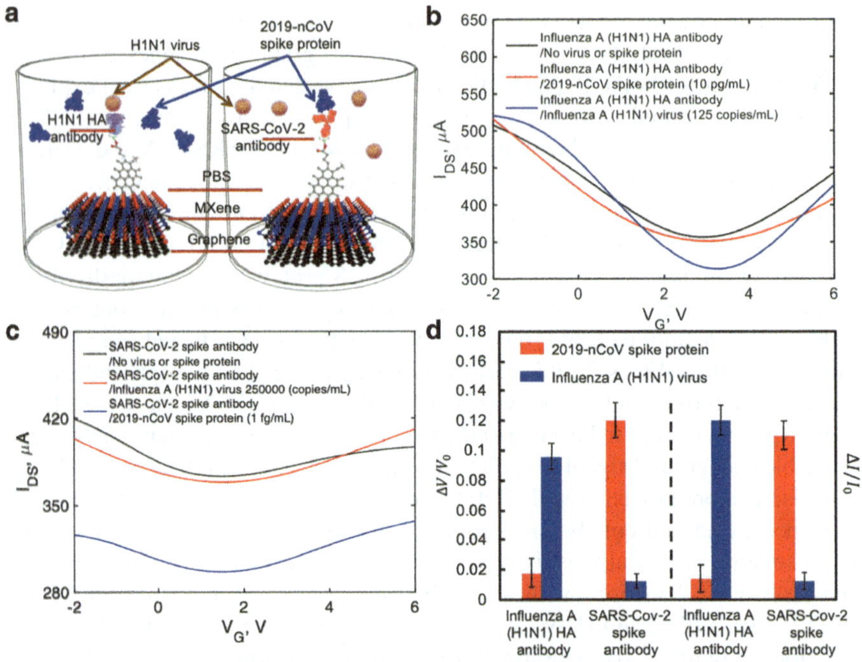

Fig. 7 **a** Illustration of specificity studies for other coronavirus detections using MXene GFET; I_{DS} versus V_{DS} curves for MXene GFET modified with **b** H1N1 influenza virus antibodies and **c** SARS-CoV-2 S protein antibodies; **d** normalized gate voltage variation for the detection of each of this virus, showing suitable specificity of the proposed device. Reprinted from reference [64] with permission from American Chemical Society

Its principles are based on the use of an Ag/AgCl$_{sat}$ as reference electrode (instead of gate one) and direct connection between drain and source terminals with the passage of electric current through graphene basal plane. Potential sweepings between drain and source terminals are made by a potentiostat (where drain is connected to CE terminal and source is connected to WE terminal) in relation to the intrinsic Ag/AgCl$_{sat}$ reference potential [18]. Thus, an external polarization of 0.197 V is applied in graphene interface by reference electrode while drain-to-source potential is swept versus this electrode. As a consequence, graphene interface is submitted to a field-effect in some extent. The electrochemical hybrid behavior could be proved when redox processes were observed for a graphene EEVD modified with ferrocene (by van der Waals interactions) submitted to hybrid I_{ds} versus V_{ds} potential sweepings [18]. It was concluded that if an electrochemical charge transfer process is presented within the potential to be swept between drain and source terminals, then it can be observed by this technique. More details are discussed in ref [18].

Another interesting aspect of EEVD devices regards the open circuit potential (OCP) of the device interface. It is understood that OCP is the potential value in which the overall electrochemical current is zero [74, 75] and reflects a condition

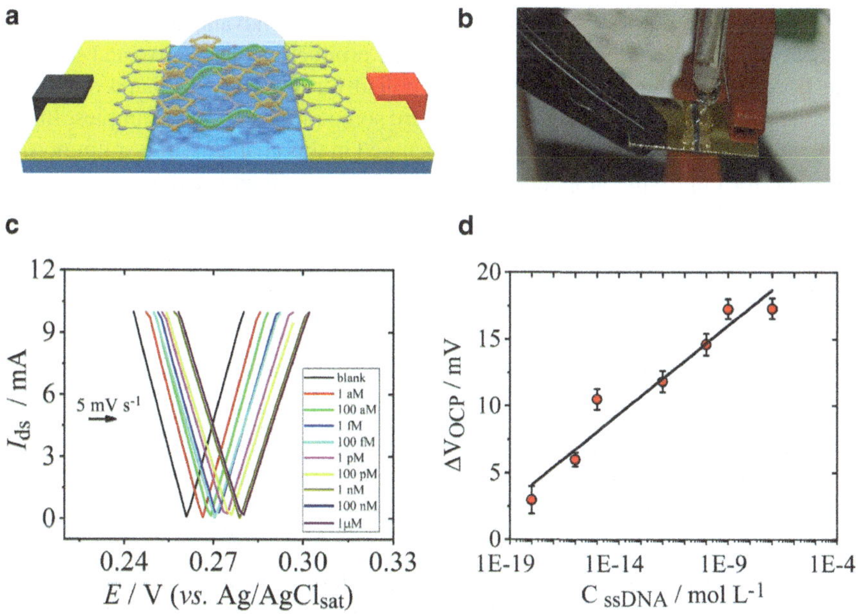

Fig. 8 **a** Schematic representation of an EEVD aiming DNA detections; **b** real photograph of an EEVD in operation; **c** I_{ds} versus V_{ds} hybrid electrical-electrochemical curves for DNA detections in a 1 amol L^{-1}—1 μmol L^{-1} concentration range; **d** respective calibration curve. Reprinted from reference [18] with permission from Elsevier

of semiconductor's electrochemistry where the sum of electron and hole charge carriers current is zero [76]. In EEVDs, it was observed that OCP value and its shifts have a similar behavior in comparison to GFET's CNP and can be used to reflect doping phenomena through charge carrier insertions [18]. It can be related to the fact that OCP has a correlation to the semiconductor interface (i.e. graphene) flatband potential and its surface charge layer. These potentials, respectively, reflect a condition where no charge excess or redox potentials are present in band edges and alterations in interfacial capacitance and surface charge layer [18]. Therefore, monitoring OCP potential means to investigate interfacial potential alterations due to capacitance changes and charge carrier insertion [18].

4 Final Considerations

In this chapter, an overview of graphene obtention, features and bioelectronics was provided. Discussing graphene intrinsic properties and how they can be applied for bioelectronics purposes was of great importance for understanding how bioelectronic devices operate and how they perform can be improved by efficiently exploring the above-mentioned properties. Besides, functionalization advantages and disadvantages were discussed aiming of directioning the production of selective and sensitive detection devices. As examples, GFETs applied for COVID-19-related diagnosis, gene therapy purposes and edge-sensors were some of the pointed examples. In this sense, the authors hope that the matters discussed in this chapter will be of important relevance for improving graphene bioelectronics research area and technologies.

References

1. Novoselov, K.S., et al.: Electric field effect in atomically thin carbon films. Science **5**(80), 1–12 (2004)
2. Macedo, L.J.A., Iost, R.M., Hassan, A., Balasubramanian, K., Crespilho, F.N.: Bioelectronics and interfaces using monolayer graphene. ChemElectroChem **6**, 31–59 (2019)
3. Balandin, A.A., et al.: Superior thermal conductivity of single-layer graphene. Nano Lett. **8**, 902–907 (2008)
4. San Roman, D., Garg, R., Cohen-Karni, T. :Bioelectronics with graphene nanostructures. APL Mater.**8** (2020)
5. Geim, A.K.: Graphene: status and prospects. Science**324**(80), 1530–1535 (2009)
6. Bolotin, K.I., et al.: Ultrahigh electron mobility in suspended graphene. Solid State Commun. **146**, 351–355 (2008)
7. Yan, Z., Peng, Z., Tour, J.M.: Chemical vapor deposition of graphene single crystals. Acc. Chem. Res. **47**, 1327–1337 (2014)
8. Mishra, N., Boeckl, J., Motta, N., Iacopi, F.: Graphene growth on silicon carbide: a review. Phys. Status Solidi Appl. Mater. Sci. **213**, 2277–2289 (2016)
9. Stanford, M.G., et al.: High-resolution laser-induced graphene. Flexible electronics beyond the visible limit. ACS Appl. Mater. Interfaces **12**, 10902–10907 (2020)
10. Lin, J., et al.: Laser-induced porous graphene films from commercial polymers. Nat. Commun. **5**, 5–12 (2014)
11. Pennacchio, F.A., Garma, L.D., Matino, L., Santoro, F.: Bioelectronics goes 3D: new trends in cell-chip interface engineering. J. Mater. Chem. B **6**, 7096–7101 (2018)
12. Cardoso, A.R., et al.: Molecularly-imprinted chloramphenicol sensor with laser-induced graphene electrodes. Biosens. Bioelectron. **124–125**, 167–175 (2019)
13. Li, J., Rogers, J.A.: Interface engineering of Si hybrid nanostructures for chemical and biological sensing. Adv. Mater. Technol. **5**, 1–13 (2020)
14. Khan, Y., et al.: Inkjet-printed flexible gold electrode arrays for bioelectronic interfaces. Adv. Funct. Mater. **26**, 1004–1013 (2016)
15. Sun, Y., Sun, M., Xie, D.: Graphene electronic devices. In: Graphene: Fabrication, Characterizations, Properties and Applications. Elsevier Inc., (2017). https://doi.org/10.1016/B978-0-12-812651-6.00005-7
16. Donnelly, M., Mao, D., Park, J., Xu, G.: Graphene field-effect transistors: the road to bioelectronics. J. Phys. D. Appl. Phys.**51** (2018)
17. Carrara, S.: What is bioelectronics? In: Handbook of Bioelectronics, pp. 1–4 Cambridge University Press (2015)
18. Mattioli, I.A., Hassan, A., Sanches, N.M., Vieira, N.C.S., Crespilho, F.N.: Highly sensitive interfaces of graphene electrical-electrochemical vertical devices for on drop atto-molar DNA detection. Biosens. Bioelectron. **175**, 112851 (2021)
19. Malard, L.M., Pimenta, M.A., Dresselhaus, G., Dresselhaus, M.S.: Raman spectroscopy in graphene. Phys. Rep. **473**, 51–87 (2009)
20. Reddy, D., Register, L.F., Carpenter, G.D., Banerjee, S.K.: Graphene field-effect transistors. J. Phys. D. Appl. Phys.**45**, (2012)

21. Ogawa, Y., et al.: Structure and transport properties of the interface between CVD-grown graphene domains. Nanoscale **6**, 7288–7294 (2014)
22. Zhu, Y., et al.: Graphene and graphene oxide: synthesis, properties, and applications. Adv. Mater. **22**, 3906–3924 (2010)
23. Huan, Q., et al.: Direct four-probe measurement of grain-boundary resistivity and mobility in millimeter-sized graphene. Nano Lett. 13–18 (2017). https://doi.org/10.1021/acs.nanolett.7b01624
24. Min, S.K., Kim, W.Y., Cho, Y., Kim, K.S.: Fast DNA sequencing with a graphene-based nanochannel device. Nat. Nanotechnol. **6**, 162–165 (2011)
25. Kalhor, N., Boden, S.A., Mizuta, H.: Sub-10 nm patterning by focused He-ion beam milling for fabrication of downscaled graphene nano devices. Microelectron. Eng. **114**, 70–77 (2014)
26. Jayasekera, T., Mintmire, J.W.: Transport in multiterminal graphene nanodevices. Nanotechnology**18** (2007)
27. Reddy, D., Register, L.F., Carpenter, G.D., Banerjee, S.K.: Erratum: graphene field-effect transistors. J. Phys. D: Appl. Phys. **44**(313001) (2011); J. Phys. D. Appl. Phys. **45** (2012)
28. Xu, H., et al.: Top-gated graphene field-effect transistors with high normalized dirac point voltage. ACS Nano **5**, 5031–5037 (2011)
29. Fu, W., et al.: Biosensing near the neutrality point of graphene. Sci. Adv. **3**, 1–8 (2017)
30. Criado, A., Melchionna, M., Marchesan, S., Prato, M.: The covalent functionalization of graphene on substrates. Angew. Chemie - Int. Ed. **54**, 10734–10750 (2015)
31. Cançado, L.G., et al.: Quantifying defects in graphene via Raman spectroscopy at different excitation energies. Nano Lett. **11**, 3190–3196 (2011)
32. Pumera, M.: Electrochemistry of graphene, graphene oxide and other graphenoids: review. Electrochem. commun. **36**, 14–18 (2013)
33. Chen, X., et al.: Ultratrace antibiotic sensing using aptamer/graphene-based field-effect transistors. Biosens. Bioelectron. **126**, 664–671 (2019)
34. Berninger, T., Bliem, C., Piccinini, E., Azzaroni, O., Knoll, W.: Cascading reaction of arginase and urease on a graphene-based FET for ultrasensitive, real-time detection of arginine. Biosens. Bioelectron. **115**, 104–110 (2018)
35. Park, D., et al.: Multiplexed femtomolar detection of Alzheimer's disease biomarkers in biofluids using a reduced graphene oxide field-effect transistor. Biosens. Bioelectron.**167**, 112505 (2020)
36. Brownson, D.A.C., Varey, S.A., Hussain, F., Haigh, S.J., Banks, C.E.: Electrochemical properties of CVD grown pristine graphene: monolayer-versus quasi-graphene. Nanoscale**6**, 1607–1621 (2014)
37. Hetemi, D., Noël, V., Pinson, J.: Grafting of diazonium salts on surfaces: application to biosensors. Biosensors**10** (2020)
38. Macedo, L.J.A., et al.: Interplay of non-uniform charge distribution on the electrochemical modification of graphene. Nanoscale **10**, 15048–15057 (2018)
39. Dong, X., et al.: The electrical properties of graphene modified by bromophenyl groups derived from a diazonium compound. Carbon N. Y. **50**, 1517–1522 (2012)
40. Cho, J., et al.: Enhanced electrical conductivity of polymer nanocomposite based on edge-selectively functionalized graphene nanoplatelets. Compos. Sci. Technol. **189**, 108001 (2020)
41. Janani, K., John Thiruvadigal, D.: Chemical functionalization and edge doping of zigzag graphene nanoribbon with L-(+)-leucine and group IB elements—A DFT study. Appl. Surf. Sci. **418**, 406–413 (2017)
42. Yadav, A., et al.: Selective electrochemical functionalization of the graphene edge. Chem. Sci. **10**, 936–942 (2019)
43. Nam, K.H., Cho, J., Yeo, H.: Thermomechanical behavior of polymer composites based on edge-selectively functionalized graphene nanosheets. Polymers (Basel).**10** (2017)
44. Liu, H., et al.: Photochemical reactivity of graphene. J. Am. Chem. Soc. **131**, 17099 17101 (2009)
45. Hassan, A., et al.: A three component-based van der Waals surface vertically designed for biomolecular recognition enhancement. Electrochim. Acta**376**, 138025 (2021)

46. Zribi, B., Haghiri-Gosnet, A. M., Bendounan, A., Ouerghi, A., Korri-Youssoufi, H.: Charge transfer and band gap opening of a ferrocene/graphene heterostructure. Carbon N. Y.**153**, 557–564 (2019)
47. Cluff, K.J., Blümel, J.: Adsorption of ferrocene on carbon nanotubes, graphene, and activated carbon. Organometallics **35**, 3939–3948 (2016)
48. Van Den Beld, W.T.E., et al.: In-situ Raman spectroscopy to elucidate the influence of adsorption in graphene electrochemistry. Sci. Rep. **7**, 1–8 (2017)
49. Tian, M., et al.: RNA detection based on graphene field-effect transistor biosensor. Adv. Condens. Matter Phys. 1–6 (2018)
50. Perumal, S., Atchudan, R., Immanuel Edison, T.N.J., Shim, J.J., Lee, Y.R.: Exfoliation and noncovalent functionalization of graphene surface with Poly-N-Vinyl-2-pyrrolidone by in situ polymerization. Molecules**26** (2021)
51. Arifin, N.F.T., et al.: Preparation and characterization of APTES-functionalized graphene oxide for CO_2 adsorption. J. Adv. Res. Fluid Mech. Therm. Sci. **61**, 297–305 (2019)
52. Shang, X., Park, C.H., Jung, G.Y., Kwak, S.K., Oh, J.H.: Highly enantioselective graphene-based chemical sensors prepared by chiral noncovalent functionalization. ACS Appl. Mater. Interfaces **10**, 36194–36201 (2018)
53. Cai, B., et al.: Gold nanoparticles-decorated graphene field-effect transistor biosensor for femtomolar MicroRNA detection. Biosens. Bioelectron. **74**, 329–334 (2015)
54. Campos, R., et al.: Attomolar label-free detection of DNA hybridization with electrolyte-gated graphene field-effect transistors. ACS Sens. **4**, 286–293 (2019)
55. Cai, B., et al.: Ultrasensitive label-free detection of PNA-DNA hybridization by reduced graphene oxide field-effect transistor biosensor. ACS Nano **8**, 2632–2638 (2014)
56. Sun, J., et al.: Magnetic graphene field-effect transistor biosensor for single-strand DNA detection. Nanoscale Res. Lett. **14**, 1–8 (2019)
57. Dontschuk, N., et al.: A graphene field-effect transistor as a molecule-specific probe of DNA nucleobases. Nat. Commun.**6** (2015)
58. Zhou, L., et al.: Novel graphene biosensor based on the functionalization of multifunctional nano-bovine serum albumin for the highly sensitive detection of cancer biomarkers. Nano-Micro Lett.**11** (2019)
59. Islam, S., et al.: A smart nanosensor for the detection of human immunodeficiency virus and associated cardiovascular and arthritis diseases using functionalized graphene-based transistors. Biosens. Bioelectron. **126**, 792–799 (2019)
60. Kim, D.J., et al.: Reduced graphene oxide field-effect transistor for label-free femtomolar protein detection. Biosens. Bioelectron. **41**, 621–626 (2013)
61. Piccinini, E., et al.: Enzyme-polyelectrolyte multilayer assemblies on reduced graphene oxide field-effect transistors for biosensing applications. Biosens. Bioelectron. **92**, 661–667 (2017)
62. Ghosh, S., Khan, N.I., Tsavalas, J.G., Song, E.: Selective detection of lysozyme biomarker utilizing large area chemical vapor deposition-grown graphene-based field-effect transistor. Front. Bioeng. Biotechnol. **6**, 1–7 (2018)
63. Li, J., et al.: Rapid and unamplified identification of COVID-19 with morpholino-modified graphene field-effect transistor nanosensor. Biosens. Bioelectron. **183** (2021)
64. Li, Y., et al.: MXene—Graphene field-effect transistor sensing of influenza virus and SARS-CoV-2. ACS Omega (2021). https://doi.org/10.1021/acsomega.0c05421
65. Seo, G., et al.: Rapid detection of COVID-19 causative virus (SARS-CoV-2) in human nasopharyngeal swab specimens using field-effect transistor-based biosensor. ACS Nano (2020). https://doi.org/10.1021/acsnano.0c02823
66. Ohno, Y., Maehashi, K., Yamashiro, Y., Matsumoto, K.: Electrolyte-gated graphene field-effect transistors for detecting pH and protein adsorption. Nano Lett. **9**, 3318–3322 (2009)
67. Tarasov, A., et al.: Gold-coated graphene field-effect transistors for quantitative analysis of protein-antibody interactions. 2D Mater. **2**, 44008 (2015)
68. Wang, X., Hao, Z., Olsen, T.R., Zhang, W., Lin, Q.: Measurements of aptamer-protein binding kinetics using graphene field-effect transistors. Nanoscale **11**, 12573–12581 (2019)

69. Khan, N.I., Song, E.: Detection of an il-6 biomarker using a gfet platform developed with a facile organic solvent-free aptamer immobilization approach. Sensors (Switzerland) **21**, 1–16 (2021)
70. Ohno, Y., Maehashi, K., Matsumoto, K.: Label-free biosensors based on aptamer-modified graphene field-effect, 18012–18013 (2010)
71. Hao, Z., et al.: Modulating the linker immobilization density on aptameric graphene field effect transistors using an electric field. ACS Sens. **5**, 2503–2513 (2020)
72. Salehirozveh, M., Dehghani, P., Zimmermann, M., Roy, V.A.L., Heidari, H.: Graphene field effect transistor biosensors based on aptamer for amyloid-β detection. IEEE Sens. J. **20**, 12488–12494 (2020)
73. Ganguli, A., et al.: High sensitivity graphene field effect transistor-based detection of DNA amplification. Adv. Funct. Mater. **2001031**, 1–9 (2020)
74. Bard, A.J., Faulkner, L.R.: Electrochemical Methods, Fundamentals and Applications. Wiley (2001). https://doi.org/10.1021/ed060pa25.1
75. Lefrou, C., Fabry, P., Poignet, J.-C.: Electrochemistry, the Basics, with Examples. Springer, Berlin, Heidelberg (2009)
76. Zhang, X.G.: Basic theories of semiconductor electrochemistry. Electrochem. Silicon Oxide (2005). https://doi.org/10.1007/0-306-47921-4_1

The manufacturer's authorised representative in the EU is Springer Nature Customer Service Centre GmbH, Europaplatz 3, 69115 Heidelberg, Germany. If you have any concerns regarding our products, please contact ProductSafety@springernature.com

Printed and bound by CPI Group (UK) Ltd, Croydon, CR0 4YY
30/09/2026
02225309-0008